T0215825

Desirable Science Education

Theresa Schulte

Desirable Science Education

Findings from a Curricular Delphi Study on Scientific Literacy in Germany

 **Springer** Spektrum

Theresa Schulte
Berlin, Germany

Dissertation Freie Universität Berlin, 2015

This work was funded by the Konrad-Adenauer-Stiftung.

OnlinePlus material to this book can be available on
http://www.springer.com/978-3-658-18254-0

ISBN 978-3-658-18253-3   ISBN 978-3-658-18254-0   (eBook)
DOI 10.1007/978-3-658-18254-0

Library of Congress Control Number: 2017941542

Springer Spektrum

Printed on acid-free paper

This Springer Spektrum imprint is published by Springer Nature
The registered company is Springer Fachmedien Wiesbaden GmbH
The registered company address is: Abraham-Lincoln-Str. 46, 65189 Wiesbaden, Germany

Summary

This study was conducted in the context of the PROFILES project as one of several national Delphi studies and part of the International PROFILES Curricular Delphi Study on Science Education. On the basis of the chosen methodology, this work represents the first approach to a modern understanding of desirable aspects of scientific literacy based science education from the perspective of different German stakeholders on a scientifically and empirically profound basis. Various stakeholders involved in science education (students at school, science teachers, science education researchers, and scientists) were asked about aspects of science education they considered meaningful and pedagogically desirable for the scientifically literate individual in the society of today and the near future. Their opinions were collected, structured, and analyzed on the basis of the Delphi method within three consecutive rounds.

In the first round, stakeholders' opinions were collected within an open response questionnaire. Specifically, they reported on situations, motives, and contexts that could facilitate science-related educational processes, meaningful contents, methods and themes of science education, and competences and attitudes that should be developed and enhanced to support students in becoming scientifically educated. Through qualitative content analysis, the open-text responses were systematized and classified into categories. In the second round, in light of the general group opinion, the stakeholders assessed categories derived from the general group opinion in terms of their priority for science education and their realization in practice. As a result, aspects considered most important for science education were identified. Moreover, by determining the difference between the provided priorities and perceived degrees of realization in current science education (priority-practice differences), areas with the most need for stronger emphasis and improvement in the science classroom were identified. In addition, stakeholders were in the second round asked to group categories that seemed meaningful to them in their combination. These combinations were analyzed through hierarchical cluster analysis in order to account for the variety of categories and determine on an empirical basis larger concepts of desirable science education. Also, they represent a more condensed picture of the stakeholders' opinions. The identified concepts were summarized as an "awareness of the sciences in current, social, globally relevant and occupational contexts relevant in both educational and out-of-school settings", "intellectual education in interdisciplinary scientific contexts", and "general science-related education and facilitation of interest in contexts of nature, everyday life and living environment". In the third round, these concepts were assessed in terms of priority and practice again, this time regarding science education in general and according to different levels of education. As a result, it was possible to identify areas in need of improvement not only in science education in general, but also at different levels of education.

As stakeholders were involved from a variety of areas affected by science education including students as the main and final users of education systems, whose views are too often neglected, a particularly comprehensive approach towards outlining meaningful scientific literacy based science education was carried out in this study. While sharing similar opinions in the majority of aspects, also several differences between groups could be identified, especially between students and adult stakeholders. This outcome at least questions the frequently invoked consensus about aspects of desirable science education and demands a more differentiated discussion of science curriculum related issues. The results of this study provide a valuable basis for such differentiated discourse.

In general, the analyses of the results show that the expectations the participating stakeholders have of scientific literacy based science education are not fulfilled in practice. This finding leads to the conclusion that among the participants, there is great discontent with the current practice of science education. In this context, the results indicate that a review of current science education practice is necessary in order to meet the aim of enhancing students' scientific literacy as part of their general education. According to the emphases by the stakeholders, science education should enhance the students' abilities of critical questioning, judgement, and reflection. In addition, science education should focus more strongly on enhancing skills and competences that are related to general aims of education, address the relation between science and society, take into account more interdisciplinary approaches and include stronger student orientation. These areas are perceived as particularly underrepresented in science education and thus seen to be in most urgent need for improvement. For all levels of education within general education, additional emphasis on references to contexts beyond scientific contents is seen as more important for meaningful science education than intellectual education in interdisciplinary scientific contexts. In addition, general science-related education and facilitation of interest in contexts of nature, everyday life and living environment should receive a stronger focus in basic education, while more advanced science education should place more emphasis on aspects related to an awareness of the sciences in current, social, globally relevant and occupational contexts.

Providing meaningful starting points for teachers to enhance students' scientific literacy, the findings of this study might serve as a fruitful source for further recommendations, inspiration, and development in the context of improving science education. On the basis of the different perspectives covered by the sample of this study, it is possible, for example, to develop empirically based learning environments that take into account the views of the students. First applications of the outcomes of this study have been administered in the context of the PROFILES project, e.g. by taking into account the results within the development of scientific literacy oriented learning and teaching materials or the preparation of professional development programs for science teachers.

How the results of this German curricular Delphi study relate to findings from other countries has already been addressed on the basis of first comparative analyses within the context of the PROFILES project. In what ways the results are replicable on an even larger international scale remains to be investigated in further research.

Zusammenfassung

Im Rahmen der gewählten Methodik dieser Studie, die als eine von mehreren nationalen Delphi-Studien und Teil der Internationalen PROFILES Curricularen Delphi-Studie Naturwissenschaften im Kontext des PROFILES Projektes durchgeführt wurde, ist es erstmals gelungen, eine wissenschaftlich und empirisch fundierte Annäherung an ein zeitgemäßes Verständnis naturwissenschaftlicher Grundbildung (scientific literacy) aus der Perspektive verschiedener Interessenvertreter (stakeholder) in Deutschland zu realisieren. Unter Einbeziehung verschiedener mit Naturwissenschaften befasster gesellschaftlicher Akteure (Schüler, Lehrer, Didaktiker und Naturwissenschaftler) wurde die Frage in den Blick genommen, welche Aspekte naturwissenschaftlicher Grundbildung für den Einzelnen in der Gesellschaft von heute und in naher Zukunft als sinnvoll und pädagogisch wünschenswert zu erachten sind. Die Ansichten dieser Interessenvertreter wurden mithilfe der Delphi-Methode im Rahmen dreier aufeinander folgenden Runden erfasst, strukturiert und analysiert.

In der ersten Befragungsrunde wurde den Teilnehmern die Möglichkeit eröffnet, im Rahmen eines offenen Fragebogens unbelastet durch inhaltliche Vorgaben ihre Vorstellungen über eine zeitgemäße naturwissenschaftliche Grundbildung auszudrücken. Diese wurden in Bezug auf Situationen, Motive und Kontexte, die zum Anlass für naturwissenschaftsbezogene Bildungsprozesse genommen werden können, Inhalte, Methoden und Themen, mit denen sich ein naturwissenschaftlich gebildeter Mensch beschäftigt haben sollte, und Fähigkeiten und Einstellungen, die das Individuum hinsichtlich der als bildungsrelevant erachteten Inhalte, Methoden und Themen erreicht haben sollte, erfasst. Die von den Teilnehmern nach dieser Einteilung formulierten Aussagen wurden anhand qualitativer Inhaltsanalyse systematisiert und in Kategorien zusammengefasst. Die durch die Kategorien repräsentierten Aspekte wurden in Kenntnis des allgemeinen Meinungsbildes im zweiten Untersuchungsabschnitt von den Beteiligten hinsichtlich ihrer Priorität für naturwissenschaftliche Grundbildung und ihrer Umsetzung in der Praxis gewichtend beurteilt. Auf Basis dieser Einschätzungen konnte ermittelt werden, welche Aspekte den Teilnehmern zufolge als am wichtigsten erachtet wurden. Auch war es auf Grundlage der Differenzen zwischen den Prioritäten der Kategorien und deren eingeschätzter Realisierung in der Praxis (Priorität-Praxis-Unterschiede) möglich, die als defizitär erachteten Bereiche naturwissenschaftlicher Grundbildung zu identifizieren. Um das durch die Kategorien repräsentierte Meinungsbild der Teilnehmer weiter zu verdichten, zielte eine weitere Aufgabe der zweiten Runde auf die Ermittlung von Konzepten wünschenswerter naturwissenschaftlicher Grundbildung. Die Akteure wurden gebeten, Kategorien zu aus ihrer Sicht sinnvollen Kombinationen zu gruppieren, die anhand einer hierarchischen Clusteranalyse in Konzepte wünschenswerter naturwissenschaftlicher Grundbildung überführt wurden. Die identifizierten Konzepte beziehen sich auf „Sensibilität für Naturwissenschaften in aktuellen, weltweit relevanten, gesellschaftlichen und beruflichen Kontexten in sowohl schulischen als auch außerschulischen Situationen", "Intellektuelle Bildung im Kontext der interdisziplinär vernetzten Naturwissenschaften", und „Allgemeine naturwissenschaftsbezogene Bildung und Interessenförderung im Kontext von Natur, Alltag und Lebenswelt". In der dritten Runde wurden diese Konzepte den Teilnehmern analog zur zweiten Runde zur gewichteten Einschätzung vorgelegt, sowohl hinsichtlich naturwissenschaftlicher Grundbildung im Allgemeinen als auch nach verschiedenen Bildungsstufen differenziert. Dadurch war es nicht nur möglich, reflektierte Rückschlüsse auf Bereiche naturwissenschaftlicher Bildung zu ziehen, in denen auf Basis der erhobenen Daten besonderer Verbesserungsbedarf besteht, sondern auch, Einblicke in spezielle Defizite bezüglich der Umsetzung der Konzepte in verschiedenen Bildungsstufen zu erlangen.

Da eine Vielzahl von Interessenvertretern aus unterschiedlichen gesellschaftlichen Bereichen, einschließlich Schülern, deren Meinung in Bildungsfragen häufig vernachlässigt oder sogar ignoriert wird, in dieser Studie beteiligt wurden, stellen die Ergebnisse der hier vorgelegten Studie einen besonders umfassenden Ansatz zur Identifizierung wünschenswerte Aspekte naturwissenschaftlicher Grundbildung dar. Trotz vielfach ähnlicher Ansichten konnten auch Unterschiede in den Meinungsbildern der verschiedenen Teilstichprobengruppen, insbesondere zwischen den Schülern und den erwachsenen Teilnehmern, aufgedeckt werden. Dieser Befund stellt den oft beschworenen Konsens über wünschenswerte naturwissenschaftliche Grundbildung zumindest in Frage und impliziert daher die Forderung nach einer differenzierteren Auseinandersetzung mit diesem Thema. Auf Basis der verschiedenen in der Stichprobe repräsentierten Perspektiven können die Ergebnisse dieser Studie eine Grundlage für einen solchen differenzierten Diskurs über wünschenswerte naturwissenschaftliche Grundbildung darstellen.

Grundsätzlich zeigen die Analysen, dass den Schwerpunkten, Bedürfnissen und Ansprüchen der beteiligten Akteure bezüglich naturwissenschaftlicher Bildung in der Praxis nicht Rechnung getragen wird und dass daher unter den Teilnehmern beträchtliche Unzufriedenheit mit der derzeitigen Praxis naturwissenschaftlicher Bildung herrscht. Insbesondere zeigen die Ergebnisse, in welchen Bereichen eine Verbesserung in der Praxis notwendig ist, damit naturwissen-

schaftlicher Unterricht seinen Anspruch, zur Allgemeinbildung der Schüler bei-
zutragen, besser als bisher erfüllt. Den Schwerpunkten der teilnehmenden Akteure zufolge sollte im naturwissenschaftlichen Unterricht allgemeinbildender Schulen insbesondere der Urteilsfähigkeit, Fähigkeit zu kritischem Hinterfragen und
Reflexion höchste Bedeutung beigemessen werden. Darüber hinaus sollte eine
naturwissenschaftliche Grundbildung stärker auf die Vermittlung von mit allge-
meineren Bildungszielen assoziierten Kompetenzen fokussieren, den Bezug
zwischen Naturwissenschaften und Gesellschaft stärken, mehr interdisziplinäre
Ansätze miteinbeziehen und mehr Schülerorientierung beinhalten. Diese Gebiete
werden auf Basis der erhobenen Daten als besonders unterrepräsentiert in naturwissenschaftlicher Bildung wahrgenommen und können daher als Bereiche mit
dem höchsten Verbesserungsbedarf angesehen werden. Ein verdichteteres Meinungsbild der Teilnehmer auf Grundlage ihrer Einschätzungen der empirisch
entwickelten Konzepte wünschenswerter naturwissenschaftlicher Bildung zeigt,
dass ein stärkerer Schwerpunkt auf Kontexten, die über naturwissenschaftliche
Fachinhalte hinausgehen, für alle vorgelegten Ebenen allgemeiner Bildung als
wichtiger für eine naturwissenschaftliche Grundbildung erachtet wird als intellektuelle Bildung im Kontext der interdisziplinär vernetzten Naturwissenschaften. Allgemeine naturwissenschaftsbezogene Bildung und Interessenförderung
im Kontext von Natur, Alltag und Lebenswelt sollte den Priorität-Praxis-
Differenzen zufolge insbesondere in der frühen und elementaren Bildung mehr
Betonung erfahren, während in den fortgeschritteneren Bildungsstufen stärkeres
Gewicht auf die Entwicklung einer Sensibilität für Naturwissenschaften in aktuellen, weltweit relevanten, gesellschaftlichen und beruflichen Kontexten gelegt
werden sollte.

Ein Hauptanliegen sollte es sein, Wege zu finden, die eine Verwirklichung
dieser Ziele in der Praxis ermöglichen. In diesem Zusammenhang können die
Ergebnisse der hier dargestellten Studie auf breiter Basis Anregungen für die
Verbesserung von naturwissenschaftlicher Bildung im Sinne einer scientific
literacy darstellen und eine Grundlage für Lehrer bieten, Schüler in der Ausfor-
mung einer naturwissenschaftlichen Grundbildung zu unterstützen. Auch ist es
aufgrund der Vielzahl an unterschiedlichen Perspektiven, die von den Interessenvertretern in dieser Studie abgedeckt sind, möglich, empirisch basierte Lern-
umgebungen zu entwickeln, die zum Beispiel die Sicht der Schülerinnen und
Schüler besonders berücksichtigen. Erste Anwendungen dieser Ergebnisse in der
Praxis wurden bereits im Kontext des PROFILES-Projektes realisiert, beispiels-
weise bei der Entwicklung von Lehr- und Lernmaterialien oder der Erstellung
von Weiterbildungsprogrammen für Lehrerinnen und Lehrer.

In welcher Beziehung die Ergebnisse dieser deutschen Studie zu Ergebnis-
sen in weiteren im PROFILES-Projekt involvierten Ländern stehen, wurde be-

reits auf der Grundlage erster Vergleichsanalysen untersucht. Inwieweit sich die Ergebnisse auf noch weiterer internationaler Ebene reproduzieren lassen, ist ein vielversprechendes Forschungsfeld für weitere anknüpfende Studien.

Contents

List of Figures

List of Tables

1 Introduction

Representing an increasingly public and policy-relevant issue across the great diversity of nations, education is a field in which research-informed findings can illuminate important concerns in society (TLRP, 2006). Especially science education is an area that is frequently discussed in terms of challenges, shortcomings and deficiencies, as indicated by recent debates on the outcomes of international comparative studies on competences of students such as PISA (OECD, 2003, 2004a, 2007b, 2010, 2014) and TIMSS (Baumert, Bos, & Lehmann, 2000; Beaton et al., 1997; Bos et al., 2008; Martin, Mullis, Foy, & Stanco, 2012). For many OECD countries, including Germany, these studies have shown that science education mostly does not equip students with the scientific knowledge and abilities they need as adults in a modern society (e.g. OECD, 2003). Also, a decline of students' interests and motivation in school science as well as large dissatisfaction with the current state of science education has been identified. In addition, it has been observed that, at least in Western democracies, science and technology are often met with distrust and suspicion (EC, 2005, 2007; Sjøberg & Schreiner, 2010). Moreover, in today's world, nations struggle to offer school education that balances the needs of society with those of the individual, academic considerations with vocational ones, and basic education with more advanced education (AAAS, 1993).

These observations have caused common concern and widespread public discussion about goals, deficiencies and prospects of science education within the frame of formal education. In many discussions, a reconsideration of the role of science education in general education and a rethinking of what is to be considered basic in science education regarding the mandate of general education is demanded. Concerns with the current state of science education are on the one hand based on the rapid development of knowledge in modern societies, the vast changes of our natural and technological environment, and the large impact of science and technology on everyday lives of individuals. On the other hand, they are also connected to the demand in scientifically and technologically sophisticated societies for qualified workforce in scientific and technical areas. Transcending national and geographic boundaries, this perceived disjuncture between science education and the realities of scientifically and technologically oriented societies has in Germany and around the world resulted in claims that underline the importance of a scientifically literate society that is prepared for meeting the demands and challenges of the twenty-first century. In this course, scientific literacy has become an issue of paramount importance in modern societies (Gräber, Nentwig, Koballa, & Evans, 2002; OECD, 2007b; UNESCO, 1993).

As science education in schools is an essential preparation for scientific literacy in later life (OECD, 2000), there is wide consensus about promoting scientific literacy as a major goal and outcome of science education within the frame of compulsory education in various countries around the world (Bybee, 2002; Eckebrecht & Schneeweiß, 2003; Fensham, 2002; Gräber & Bolte, 1997; Gräber et al., 2002; Hurd, 1998; NRC, 1996; Walberg & Paik, 1997). In response to these claims, reforming the curriculum towards achieving scientific literacy for all students has been and is a major undertaking in Europe and beyond (AAAS, 2001; EC, 2011). Hence, it is not surprising that scientific literacy has in recent years been included as a central aim of science education in science curricula of compulsory education as well as the national science education standards in Germany (KMK, 2005a, 2005b, 2005c).

Even though scientific literacy has been increasingly used in recent years to characterize the central aim of school science education, many of the claims and recommendations subsumed in the literature under this catchword are more descriptive than research-based. Also, the large variety of denotations associated with this concept suggests a certain vagueness about its particular meaning and implications for the science curriculum (Bybee, 2002; Coll & Taylor, 2009; DeBoer, 2000; Gräber & Bolte, 1997; Gräber & Nentwig, 2002; Koballa, Kemp, & Evans, 1997; Laugksch, 2000; Shamos, 1995). While comparative studies such as PISA or TIMSS have made a contribution to defining competences that can be developed in science education, they can however, as part of evaluative research that has defined and measured wanted outcomes, not provide answers to the question in which ways the objectives of such literacy can be achieved, but only offer starting points for science curriculum related discussions (Fensham, 2007, p. 117). Also, as in most societies, aspects that are both important and salient within a given domain such as science education are usually defined by the academic community (Osborne, Ratcliffe, Collins, Millar, & Duschl, 2003), which inevitably suggests that the voices of students, teachers, scientists, or other relevant stakeholders in society are often suppressed when it comes to the question about what is important for young people to know, value, and be able to do in situations involving science and technology as scientifically literate citizens of tomorrow.

In view of these considerations, the question arises how solid and applicable the frequently invoked consensus in the scientific community on facets of meaningful science education in the sense of scientific literacy is. To find answers to this question, it seems necessary to directly involve different stakeholders in critical reflection and participatory research with regard to curricular issues in science education. For these reasons, this study aims at a reconsideration of contents, methods, and aims of scientific literacy based science education from the perspective of different German stakeholders. Within this context, it sets out to conduct a research-based approach to providing a framework for modern science education with the aim of enhancing students' scientific literacy. To accomplish this task, various stakeholders involved with science education in Germany (students at school, science teachers, science education researchers, and scientists) are asked about what aspects of science education they consider meaningful and pedagogically desirable for the scientifically literate individual in the society of today and the near future. For the purpose of collecting, structuring, and analyzing these stakeholders' views within this study, the Delphi method is chosen.

This method has previously proven suitable for identifying aspects considered relevant and desirable in different subjects and issues of science education (Bolte, 2003a, 2003b, 2008; Häußler, Frey, Hoffmann, Rost, & Spada, 1980; J. Mayer, 1992; Osborne et al., 2003). Based on the proposed procedure by Bolte (2003b), this study applies three consecutive rounds in which stakeholders first contribute their views about aspects of desirable science education and then assess the compiled aspects and results of each round with respect to more specific evaluations. While the aforementioned previous studies were conducted for specific issues within science education or particular science disciplines only, this study includes a more comprehensive approach with a focus on science education as a representation of all science subjects. Moreover, this study is carried out in the context of the PROFILES project (PROFILES, 2010b) as part of the International PROFILES Curricular Delphi Study on Science Education and as one of several national Delphi studies conducted in different European countries. As the PROFILES project brings together a large number of stakeholders involved with science education and science from different countries, this frame also holds out the prospect of insightful follow-up investigations on an international scale.

As a basis for the analyses, I will in the theoretical part of this study first address the role of science within general education and reflect on goals of science education on the basis of an overarching concept of general education. In particular, I will discuss the relevance of science education for the individual and society with reference to the concept of scientific literacy. Also, I will introduce the PROFILES project as the frame in which this study is conducted. In addition, I will illustrate the curricular elements of science education that are examined in this investigation and discuss central components of curricular processes and relevant stakeholders of curricular processes in science education. Moreover, I will describe the Delphi method as a systematic approach to involving a wide range of stakeholders and bringing together their views and opinions as a suitable tool for the investigation of curricular aspects of science education. In order to gain further insights for both the conception of this study and methodological aspects of the analyses, I will illustrate the current state of research on views about meaningful science education on the basis of the results of previous subject-specific Delphi studies.

In the methodical part, I will describe the conception of this study, the development of the instruments of data collection, and the procedure of data collection as well as methods of data analysis.

The presentation of the results is structured according to the three rounds. For all rounds, I will provide a description of the sample and the response behavior. With respect to the outcomes of the first round, I will present the inter-rater agreement, the classification system, and a characterization of the responses, as well as descriptive statistical analyses such as category frequencies. I will illustrate the findings from the second round in terms of the stakeholders' priority and practice assessments of categories as well as resulting priority-practice differences. Moreover, I will describe the results of hierarchical clustering of category combinations. Regarding the third round, I will present the stakeholders' priority and practice assessments of concepts of science education as well as resulting priority-practice differences both for general science education and differentiated according to different levels of education. For clarity reasons, the presentation of the results of each round is immediately followed by a discussion of the corresponding results in view of the research questions as well as with reference to findings of other investigations.

In the final discussion, I will review the overall results of this study in light of the obtained interrelated factors identified to inform the operationalization of scientific literacy in practice and reflect on the results in terms of their significance for future science education in Germany. Moreover, I will compare the outcomes of this study to further national curricular Delphi studies conducted in the context of the PROFILES project, and point out implications for science education in Europe. Finally, after a reflection on limitations of this study and opportunities for further research, the results of this investigation are summarized.

2 Theoretical Framework

In the first chapter of this theoretical framework, I will address the role of science within education in general. It provides the theoretical foundations for the analyses conducted in this study in light of the question what aspects of science education are considered important for the scientifically literate individual in the society of today and in the near future. By defining related terms and concepts, I will further point out the significance of science education for the individual and society and shed light on purposes and forms of organization of the compulsory years of school science education. Moreover, I will introduce the PROFILES Project both as an example of efforts to enhance science education in Europe and as the framework in which this study has been carried out. In the second chapter, I will discuss the theory of curriculum and curricular dimensions of science education with respect to science education as a curriculum-based, institutionalized endeavour. In the third chapter, I will describe elements of a curricular process and identify central stakeholders of society that are affected by science education. Furthermore, I will introduce the Delphi method as a means to access, collect and structure opinions of a group of stakeholders within a curricular process. In the fourth chapter, I will provide insights into findings of previous science curriculum related Delphi studies in physics, biology, and chemistry. In the fifth chapter, I will present the research questions and hypotheses of this study.

2.1 Science in the Context of General Education

There is widespread recognition that science is an essential part of education (e.g. Gräber & Nentwig, 2002; Hollanders & Soete, 2010; OECD, 1999; UNESCO, 1993). When considering science education in Germany within the broader context of education in general and addressing the importance of science for general education with respect to the overarching ideas of education, some reflections have to be made on concepts of general education with an emphasis on the German theory of education and the German concept of *Allgemeinbildung*. Fundamental to addressing the role of science within mandatory school education is also a clarification of the concept of scientific literacy and a discussion of the purposes of scientific literacy based science education. Moreover, as this study is part of a European endeavor which approaches science education from a holistic perspective, I will discuss the rationale for an integrated perspec-

tive on science in the context of education and introduce the PROFILES Project
is introduced both as an effort to enhance science education in Europe and as the
framework in which this endeavor is embedded.

2.1.1 Reflections on Concepts of General Education

Education can be seen as one of the central themes in society (Tenorth, 1986a).
General education in terms of students' personal development and their preparedness as citizens in modern society is often linked to the unique German concept of *Allgemeinbildung*, which can very roughly be translated as "general education" or "education for all". The idea of *Allgemeinbildung* traces back to the
roots of the European educational tradition and refers to the overarching purposes of education. Having been influenced by different thoughts of educational
theory over the time, *Allgemeinbildung* represents a multi-faceted construct with
respect to its understanding and interpretation (Benner, 2002; Heymann, van
Lück, Meyer, Schulze, & Tenorth, 1990; Klafki, 2000, 2007; Sühl-Strohmenger,
1984; Tenorth, 1986b, 1994, 2003, 2006). In order to shed light on the concept of
Allgemeinbildung, the following subchapter will clarify the underlying ideas and
related terms of this concept.

Within *Allgemeinbildung*, the two terms *allgemein* and *Bildung* can be distinguished. The noun *Bildung*[1] is the term for a key idea in the German theory of
education. As there is no precise English translation and a simple translation as
education might invoke misleading connotations, the concept of *Bildung* is in
English research literature also denoted by paraphrases such as formative education (Vásquez-Levy, 2002), liberal education (Løvlie & Standish, 2002; Westbury, 1995, p. 243), cultivation (Westbury, 1995, p. 234) or as formation of personality through education (Westbury, 2000). Vásquez-Levy (2002, p. 118) underlines that "English must employ several different articulations in order to
capture the many aspects of the German expression". She points out that when
focusing on the developmental aspects of an individual within a discussion on
Bildung, one speaks of maturation, whereas the individual's relation to his or her
environment are addressed in terms of emancipation. Further facets of meaning
are social aspects such as "the capacity to open up to diverse circumstances to
extend the concept of humanity through responsible activity" (Vásquez-Levy,
2002, p. 118). With respect to translating the role of education in terms of *Bild-*

---

1 The discussion of the German term Bildung in this place is meant to consider its meaning and
  implications with respect to addressing contemporary science education. A detailed elaboration is
  beyond the scope of this work. Comprehensive accounts are e.g. provided by Klafki (2000, 2007),
  Tenorth (1986a, 1994, 2003, 2006), von Hentig (1996), and Blankertz (1984).

ung into English, expressions such as the "the ideal of self-determination, the formation of character, or exercise of autonomy, reason and independence" are used (Vásquez-Levy, 2002, p. 118). For an approach to the idea of *Bildung* in English, it has thus to be noted that "[a]ll of these labels and qualities in their English forms are inherent in the German term *Bildung"* (Vásquez-Levy, 2002, p. 118). Therefore, definitions of the term in the English-speaking research community with reference to the variety of facets of the German idea of *Bildung* include descriptions such as "the process of developing a critical consciousness and of character-formation, self-discovery, knowledge in the form of contemplation or insight, an engagement with questions of truth, value and meaning (Vásquez-Levy, 2002, p. 118)". The emancipatory sense of *Bildung* is captured in the meaning of the German verb *bilden* (which can be translated as "to form" or "to shape") (Tenorth, 2006, p. 7). This connotation emphasizes the process of an individual's personal development and is in this way often also related to the concept of action competence (Mogensen & Schnack, 2010) and citizenship (Elmose & Roth, 2005). As the concept of *Bildung* in its original sense is understood in a broader sense than English translations, the German term is used in international research literature as well (Elmose & Roth, 2005; Nijhof, 1990; Sjöström, 2013; Wimmer, 2003).

Bildung and its interpretation as *Allgemeinbildung* became a central focus of reflection around the transition from the 18th to the 19th century, a time that was characterized by the later part of the Enlightenment, philosophical-educational idealism, the classical period of German literature, neo-humanism, and certain undercurrents of Romanticism (Klafki, 2000, p. 85). As the concept of *Bildung* has throughout time been subject to different thoughts and influences of society and cultural contexts, *Bildung* is a dynamic construct that has undergone a variety of shifts with respect to its meaning and interpretation (Klafki, 2000; Tenorth, 1986a). The political-moral connotation of *Bildung*, for example, faded at the end of the 19th century, as *Bildung* became increasingly associated with aesthetic education (Pinar, 2009, p. 27).

With different approaches to a contemporary use of the concept, *Bildung* features a multitude of differentiations. Yet, since the 1970s, *Bildung* has become an increasingly discussed subject of social relevance (Blankertz, 1984; Heymann et al., 1990; Klafki, 2007; Klemm, Rolff, & Tillmann, 1985; Neuner, 1999; OECD, 1999; Sühl-Strohmenger, 1984; Tenorth, 1986a; von Hentig, 1996; Wimmer, 2003). In this context, *Bildung* contains as two major pillars a personal and a social dimension. According to this division, *Bildung* should on the one hand contribute to individual self-development, on the other hand, *Bildung* is also realized in a social context and related to the notion of social responsibility (J. Mayer, 1992, p. 14).

One of the most influential and distinguished scholars with respect to a contemporary understanding of the concept of *Bildung* is the German scholar Wolfgang Klafki, whose works had great impact on educational theory in Germany and beyond. He underlines that in German, *Bildung* means both the process and the product of becoming educated *(gebildet)* with the help of others (Klafki, 1995a, p. 15). Based on the idea of mutual interrelatedness between the individual and the world, Klafki (2000, pp. 91ff.) proposes a dialectical understanding of *Bildung*. He describes *Bildung* in its double meaning as both the world being accessible to the individual (the objective or material aspect) and the individual being accessible to the world (the subjective or formal aspect), which leaves *Bildung* between the two poles of the needs and interests of the individual and the requirements of society. This inextricable link between subjectivity and objectivity in *Bildung* takes place in the relation of the self to the world (Klafki, 2000, p. 91). Hence, according to Klafki (2000, p. 93) self-formation occurs through an engagement with the world so that the cultivation of personal uniqueness and individuality does not occur in isolation but only in communication with others through processes of recognition and mediation. Klafki (1964) links these two components in a categorical concept of *Bildung* that is characterized by the principles of the elementary, the fundamental and the exemplary. For Klafki (2000, pp. 87ff.), an essential feature of *Bildung* is helping students to develop three basic abilities that trace back to the tradition of German education theory: the competences for self-determination, participation in society, and solidarity with others. Moreover, he names responsibility, reason, emancipation, and independence as abilities to achieve through *Bildung*. Describing *Bildung* as embracing all dimensions of human abilities, Klafki's understanding of *Bildung* includes attitudes and skills such as aesthetic awareness, acting in society, the ability to judge and to organize, self-reflection, argumentation skills, empathy, complex reasoning, and a reflective relation towards others and the world a result of the educational process. In this way, *Bildung* can be seen as a mediator between the individual and the world.

Taking into account these considerations, it becomes apparent that the idea of *Bildung* cannot be reduced to fostering reason and rationality. With the notion that society needs educated citizens who are able to make informed decisions (Fleming, 1989; Gräber et al., 2002), the contemporary use of the concept is shaped by the notions that *Bildung* should serve to prepare the students for effective participation in a technologically sophisticated society by enabling them to become responsible and reflective citizens (Fleming, 1989; OECD, 2000). Moreover, the current notion of *Bildung* refers to being part of the constant process of the students' personal formation and socialization throughout the course of which several overarching abilities are enhanced (Klafki, 2000, pp. 85ff.). In general, it can thus be said that *Bildung* has "as its aim the fulfillment of humanity: full development of the capacities and powers of each human individual to questioned preconceived opinions, prejudices, and 'given facts', and intentioned participation in the shaping of one's own and joint living conditions" (Mogensen & Schnack, 2010, p. 61).

Tying in with the notions of the German tradition of *Bildung*, Bolte (2008, p. 332) defines *Bildung* as

"the idea, necessity, task, process, and endeavour to form one's own identity and enlightened world view in a self-determined examination of the world; to gain knowledge and abilities in order to find orientation as well as to become capable of acting and judging".

However, Bolte indicates that self-regulated personality development as rooted in the classical notion of *Bildung* is limited within school education and thus points to Tenorth's understanding of the concept of *Allgemeinbildung* as

"[...] all efforts of a society, culture, or nation that serve, by means of societal institutions, to spread that knowledge and those abilities and attitudes among the adolescent generation whose mastery is historically regarded as being necessary and indispensable" (Tenorth, 1994, p. 7, translation following Bolte, 2008, p. 332).

As these efforts are not determined by the students, it can be argued that *Allgemeinbildung* as the central task of school education (Eckebrecht & Schneeweiß, 2003; Heymann, 1990; Schecker et al., 1996; Sühl-Strohmenger, 1984) constitutes a shift from the subjective focus of *Bildung* to *Bildung* being seen as a bundle of associated tasks featuring consensus in contemporary society, and which point to a more functional meaning of *Bildung* (Heymann et al., 1990; Schulze, 1990). In fact, it can be noted that today, there is a dominant understanding of *Bildung*

"as a central element and instrument to equip the individual with relevant knowledge, competences and skills to cope with the dynamism of societal change and expectations" (Wimmer, 2003, p. 168).

However, concerns are expressed with regard to the understanding of *Bildung* as being solely the attainment of skills and knowledge as a means to promote one's own interests in global competition (Wimmer, 2003, p. 169).

Klafki (2000, pp. 88ff.) describes the efforts of general education in terms of *Allgemeinbildung* by distinguishing three shades of meaning of the prefix *allgemein*. The first refers to a *Bildung* that is addressed to everyone. The second facet relates to a *Bildung* that takes place in the medium of a general, that is, "of historical objectifications of humanity, of humaneness and its conditions, with an orientation to the possibilities of, and obligation to, humanitarian progress"

(Klafki, 2000, p. 92). By this, Klafki means engaging with common key issues featuring cultural, social, political and personal dimensions. The third aspect refers to the individual's development of versatility embracing all dimensions of human interest, such as cognitive, social and emotional aspects and going beyond the classical canon of education. In this way, *Allgemeinbildung* is not seen as a mosaic of specialized knowledge of academic subjects but rather defined as *Bildung* in a threefold sense.

Relating Tenorth's and Klafki's definitions of *Allgemeinbildung* and referring their understanding of *Allgemeinbildung* to science education, science related *Allgemeinbildung* can be defined from the viewpoint of the German theory of education according to Bolte (2008, p. 332) from the perspective of science as

"those efforts of general education, which are addressed to all people, contribute to the individual's formation or versatility, and should be promoted in the framework of general global problems"

As the process of *Allgemeinbildung* is closely linked to and fostered through content, the concrete tasks of the different realms of formal education should be reflected in the light of the overarching aims of education (Heymann et al., 1990; Schecker et al., 1996; Tenorth, 1994). According to the theory of *Bildungsgangdidaktik*, which especially takes into account the educational process of the individual (e.g. Meyer & Reinartz, 1998; Schenk, 2005), educational content should in particular refer to developmental tasks (Havighurst, 1981) such as personally relevant challenges of individuals with which they are confronted throughout their lives and within which they develop their personalities.

The authors of a report on how to increase the efficiency of mathematics and science teaching in Germany (Bund-Länder Kommission für Bildungsplanung und Forschungsförderung, 1997) point out that in the German discussion on a modern understanding of *Allgemeinbildung*, attempts have been made to reconcile the pragmatic approach of coping with life by Robinsohn with the values and personality oriented understanding of education by Wilhelm von Humboldt (Gräber, 2002, p.7). In that manner, *Allgemeinbildung* is in particular understood as transcending disciplines, leading to transdisciplinary general competences independent of the particular subjects. The authors emphasize the importance of such general competences for coping with life in present and future, and highlight the basis of these competences for lifelong learning. They identify four areas of general competences (Gräber, 2002, p.7):

- mastery of basic cultural tools such as language and mathematical symbols and routines,
- contextual knowledge in key domains of knowledge,

- metacognitive competences and motivational orientation,
- social-cognitive and social skills.

With respect to educational content that is oriented towards the concept of *Allgemeinbildung*, Tenorth (1994, p. 173) formulates four dimensions featuring consensus for an *Allgemeinbildung* oriented core curriculum: language, the historical-social learning field, the mathematic-scientific learning field, and the aesthetic-expressive learning field. He continues to point out that these learning areas allow for recognizing the world as a communicative unit, to communicate within this world, and to see the problems of the world both as historically grown and as socially processable. At the same time, these structures allow for formulating one's own subjectivity and address reality beyond traditional school subjects.

The following sub-chapter discusses how these considerations relate to the role of science within general education in particular and how science education is embedded in the framework of general education.

2.1.2 The Contribution of Science to General Education

Science shapes our modern culture, impacts our individual and collective lives, and provides both theoretical visions and tangible options to improve human conditions (Ramsey, 1997, p. 325). Thus, it is widely acknowledged that science, independently of cultural or social settings, is one of the fundamental and indispensable dimensions of education in general (DeBoer, 1991; Tenorth, 1994).

The relationship between science and education can be traced back to the introduction of modern science into Western civilization at the end of the 16th century (Hurd, 1998, p. 407) and its advancement during the 17th century within the establishment of natural philosophy as a social institution (Aikenhead, 2003, p. 4). Notwithstanding the classical and humanist educational traditions, notions developed that general education in the sense of *Allgemeinbildung* also takes place and is achieved through engaging with science. It was further emphasized that the study of science serves the understanding of the modern world and that science education should be connected to the student's real world experiences (von Engelhardt, 2010). Special influence on the rationale of science as a part of general education is attributed to thinkers of the Age of Enlightenment such as Francis Bacon (Schöler, 1970, p. 23) who stated in 1620 that

> "the ideal of human service is the ultimate goal of scientific effort, to the end of equipping the intellect for a better and more perfect use of human reason" (Dick, 1955, p. 441 quoted in Hurd, 1998, p. 407).

This demand for science education grounded on the principals of empirical realism also prompted John Amos Comenius in the 17th century to promote the idea of pursuing formal educational goals through engaging with natural phenomena and to put forward an appreciation of science education and efforts towards more equal treatment of the educational value of science among the other traditional subjects (Schöler, 1970, p. 23). Theodor Litt (1959) and Martin Wagenschein (1968, 1980) were pioneers among German speaking thinkers addressing the significance of science for general education. Wagenschein underlined the necessity of science-related *Allgemeinbildung* for those parts of society that do not pursue careers in the field of science (Wagenschein, 1980, p. 11). In this way, humanistic perspectives entered the science curriculum and were described in terms of "values, the nature of science, social aspects of science, and the human character of science revealed through its sociology, history, and philosophy" (Aikenhead, 2003, p. 2).

In Germany, science education first received validity with respect to general institutionalized education and became a regular part of the curriculum of schools of general education with the establishment of the *Realschule* at the beginning of the 18th century (Bonnekoh, 1992, p. 71). A further influential factor for the development of science education in Germany was a greater acceptance of science in the public in the course of economic advancement during the 19th century (Schöler, 1970, p. 72). Thus, towards the end of the 19th century, science was increasingly seen as a fundamental aspect of culture and established as an essential part of the curriculum at schools of general education, both in Germany and beyond (Bonnekoh, 1992, p. 7; DeBoer, 1991, p. 216).

In 1902, Smith defined five potential contributions of science teaching to general education (DeBoer, 1991, p. 54):

1. Training in the powers of observation of the natural world
2. Training in a powerful method of generating new knowledge that is based on observation and experiment
3. Exercise of the imagination and creative impulses
4. Training to view problems objectively
5. Generation of useful information

However, "in the first half of the 20th century, the study of science was linked to effective living in an increasingly industrialized world" (DeBoer, 1991, p. 217). In this manner, the content of science education in school was still mainly dominated by the need to provide the foundations for professional training of future scientists for much of the last century (OECD, 2003, p. 92). This changed with the growing role of science in modern life so that science education was promoted more and more on the basis of its relevance to contemporary life (DeBoer, 1997, p. 73) and "its contribution to a shared understanding of the world on the part of all members of society" (DeBoer, 2000, p. 583). This development was also inspired by early 19th century thinkers such as John Dewey who emphasized that dealing with science enhanced the ability of critical thinking and a scientific attitude embracing aspects such as analyzing, formulating research questions, logical thinking, and relying on solid evidence (Shamos, 2002, p. 45). Dewey described science with respect to its importance for society "as a legitimate intellectual study on the basis of the power it gave individuals to act independently" (DeBoer, 2000, p. 583) and, based on "its practical applications and its focus on the real activities of life" (DeBoer, 1997, p. 72), as a potential link between science teaching and responsible citizenship in democratic societies (Oliver, Jackson, & Chun, 2001, p. 11), pronouncing that

> "[w]hatever natural science may be for the specialist, for educational purposes it is knowledge of the conditions of human action" (Dewey, 1916, p. 228 quoted in DeBoer, 2000, p. 583).

Today, it is widely recognized that scientific competences and knowledge are not just important for contemporary industrial and knowledge-based societies with respect to social progress and in order to compete with scientifically and technologically accomplished professionals on a global market. With the prevalence of the sciences in all realms of life and with every individual's responsibility in the society, a profound science education is also essential for every individual with respect to opinion-making and decision-making skills regarding one's own actions. Moreover, science is counted among the great achievements of modern society, in which a scientific understanding plays an essential role for developing an enlightened world view. Furthermore, with the development of an informed citizenry being one of the main functions of school education, it is indispensable that citizens are able to recognize and understand social issues with scientific references in order to participate in social discourse and policy-making processes (Gräber & Bolte, 1997; Gräber & Nentwig, 2002; Kolstoe, 2000; NRC, 1996; OECD, 2003).

The centrality of science related issues for an individual's education is also reflected in the consideration of scientific literacy (see 2.1.3) as a general ability for life (OECD, 2003, p. 21). Addressing science education within mandatory education, the expert group of PISA underline the significance of science for an individual's acting and managing in everyday life with regard to issues of health, disease, nutrition, sustainability, climate, environment, and consumer behavior (OECD, 2000, p. 79). Likewise, the American Association for the Advancement of Science stresses the task of schools to foster healthy, socially responsible behavior among young people within the context of science education and to prepare them for citizenship, for work, and for coping with everyday life (AAAS, 1990, 1993, 2001). For these reasons, science is seen as a fundamental dimension of general education in school (OECD, 2007b). Therefore, science education must provide a context for general educational goals, providing a basis for meaningful, generalizable learning outcomes for the students as citizens (Ramsey, 1997, p. 325).

Bridging the gap to Klafki's understanding of *Allgemeinbildung* in the sense of *Bildung* for everyone and to the growing role of science in modern life, "the objectives of personal fulfillment, employment and full participation in society" increasingly require that all adults should be scientifically literate, not just those aspiring to a scientific career (OECD, 2003, p. 92). Hence, the 'Science for All' slogan now has global resonance (Jenkins, 1999) and the necessity is underlined for a science education for all that embraces skills of opinion-making, responsibility, critical assessment and acting with regard to nature and technology (AAAS, 1990; Gräber & Bolte, 1997; Gräber et al., 2002).

Dimensions of science through which education in terms of personal formation can be achieved with particular reference to physics education are discussed by Lauterbach (1992a). Recognizing the implications of Klafki's understanding of *Bildung* for physics education, he defines as relevant epistemological dimensions education through scientific knowledge (knowledge about the world), education through knowledge of the theory of science (limits of scientific knowledge and inquiry), education through historical knowledge (social determination), education through self-awareness (construction of reality), education through discursive inquiry (validity as an application problem), and education through impacting knowledge (interaction as dialogue) (Lauterbach, 1992a, pp. 19–33).

The notion of undergoing and achieving education through a preoccupation with science with the education-through-science approach is also a central aspect of the PROFILES Project (see 2.1.5), an FP7 funded European project that aims at enhancing science education by making efforts to disseminate a modern understanding of science teaching, encouraging new approaches into the practice of science teaching and facilitating an uptake of inquiry-based science education (Bolte et al., 2011; Bolte, Holbrook, & Rauch, 2012; Bolte & Streller, 2013; PROFILES, 2010b). Following the constructivist notion of learning as a process in which the individual is actively involved, and thus implying certain ties with Klafki's education theory (Klafki, 2007) and the theory of *Bildungsgangdidaktik* (e.g. Schenk, 2005), the education-through-science approach promotes a focus on the students and proposes that education be the major emphasis, independently of being undertaken within science teaching or teaching in any other discipline. In this way, the students' needs, interests and perspectives are given precedence over subject-propaedeutic demands. Within the PROFILES framework, the education-through-science approach embraces a wide range of educational attributes, such as an appreciation of the important role science plays within the world, recognition of the relevance of education-through-science for lifelong learning, responsible citizenry, and preparing for a profession (Bolte, Streller, et al., 2012, pp. 31–33; Bolte & Streller, 2013, pp. 180–181).

In Germany, the contribution of science to general education is in part of the national education standards specified for the three science subjects biology, chemistry, and physics (KMK, 2005a, 2005b, 2005c). The particular contribution of biology education is related to an involvement with the living world. Living nature is classified and represented by various systems, e.g. cell, organism, ecosystem, biosphere, their interactions and their evolutionary history. An understanding of biological systems requires dynamic thinking and taking different perspectives. This makes it possible to develop multi-perspective and systematic thinking through biology teaching. Moreover, the human being in this system structure is part and counterpart of nature. With the human being as a subject in biology classes, biology teaching contributes to the development of individual self-awareness and emancipatory action. This is the basis for health-conscious and environmentally sustainable action both in individual and in social responsibility (KMK, 2005a, p. 6).

The contribution of an engagement with chemistry to general education lies in the unique properties of chemistry to examine and describe the material world with special reference to chemical reactions as a sum of mass and energy transformation by particle and structural changes and the modification of chemical bonds. In this way, chemistry provides insights into the composition and synthesis of substances and materials, and for dealing with them appropriately. Also, an occupation with chemistry enables students to explain and evaluate phenomena of their environment on the basis of their knowledge of substances and chemical reactions, to make decisions and judgments and to communicate them appropriately. Moreover, dealing with chemistry helps the students to recognize the importance of chemistry as a science, the chemical industry and chemistry-related professions for society, the economy and the environment. At the same time, chemistry makes the students aware of a sustainable use of resources. This includes the responsible use of chemicals and devices in household, laboratory and environment, and safety-conscious experimentation. In addition, the students learn to use the experimental method in particular as a means for individual knowledge acquisition about chemical phenomena and learn about the limits of scientific inquiry (KMK, 2005b, p. 6).

Physics serves as an essential basis for understanding natural phenomena and for the explanation and evaluation of technical systems and developments. With the unique contents and methods of physics, an occupation with physics enhances the subject-specific approaches to tasks and problems as well as the development of a specific worldview. In particular, dealing with physics enables an encounter with the world through the modeling of natural and technical phenomena and predicting the outcomes of cause-effect relationships. In this regard, both a structured and formalized description of phenomena and addressing their essential physical properties and parameters play a role. Thus, by engaging with physics, the students are provided with occasions to use the physical modeling of natural phenomena for explanations. In this way, physics education can provide a basis for young people for an occupation with scientific topics and their social contexts (KMK, 2005c, p. 6). The AAAS pronounces that as the

> "life-enhancing potential of science and technology cannot be realized unless the public in general comes to understand science, mathematics, and technology and to acquire scientific habits of mind; without a scientifically literate population, the outlook for a better world is not promising" (AAAS, 1990).

Hence, today – in Europe and around the world – science is an established part of the curricula of schools of general education with the overall aim of enhancing the students' scientific literacy as part of their general education and serving as a basis for lifelong engagement with science (e.g. OECD, 2007b).

This overall aim of science education to enhance students' scientific literacy is widely agreed on (Bybee, 1997; Eckebrecht & Schneeweiß, 2003; Gräber & Nentwig, 2002; NRC, 1996; OECD, 2000). The term scientific literacy can be translated into German as *naturwissenschaftliche Grundbildung* (Baumert et al., 1999; Gräber & Bolte, 1997; Gräber & Nentwig, 2002; OECD, 2007a; Rost, Senkbeil, Walter, Carstensen, & Prenzel, 2005). *Grundbildung* is a term used in the German-speaking scientific research community that underlines the basic and versatile facet of *Bildung* (e.g. Baumert et al., 1999, 2000, 2000; Gräber & Nentwig, 2002; M. Kremer, 2012; Lauterbach, 1993; Rost et al., 2005). In a more narrow sense, *Grundbildung* can also be attributed to school education at the primary and lower secondary level, referring in this way to those levels usually included in compulsory education (Schulze, 1990).

Within the framework of PISA, three different dimensions of literacy in the sense of *Grundbildung* are operationalized – reading literacy, mathematical literacy and scientific literacy. Similar to the overarching ideas of education, the PISA domains of literacy focus on "the ability to undertake a number of fundamental processes in a range of situations, backed by a broad understanding of key concepts, rather than the possession of specific knowledge" (OECD, 2000, p.

7). In this way, PISA frames the idea of literacy in terms of *Grundbildung* as "the ability to stand apart from arguments, evidence or text, to reflect on these, and to evaluate and criticize claims made" (OECD, 2000, p. 12) and as

"the capacity of students to apply knowledge and skills in key subject areas and to analyze, reason and communicate effectively as they pose, solve and interpret problems in a variety of situations" (OECD, 2007b, p. 16).

According to this understanding, *Grundbildung* includes facets of analysis, problem solving and communication as well as aspects of evaluation and critical reflection.

The relation of science education to such an understanding of *Grundbildung* is captured in the concept of scientific literacy as the main goal of science related general education. The following chapter addresses the thoughts that the construct of scientific literacy has come to represent and the attempts to define the meaning of scientific literacy as an educational goal within science education.

2.1.3 Scientific Literacy

As scientific literacy represents an essential component of the modern understanding of the aims and objectives of science education (Laugksch, 2000, p. 71), this chapter takes a closer look at the concept of scientific literacy.

The term scientific literacy was introduced by Conant in the 1950s and traces back to a movement in the United States, which, prompted by the postwar drive for industrialization, the Sputnik embarrassment and general dissatisfaction with the outcomes of science education, aimed at a modernization of science education and a reform of the science curriculum to enhance student achievement in science to prepare for more workforce in science and science related professions (Bybee, 1997, 2002; Hurd, 1998). Subsequently, the term was in particular brought into discussion with reference to the need to educate citizens who are able to understand scientific contributions and engage in meaningful democratic participation, representing a step toward building a civic dimension of scientific literacy (Hurd, 1998, p. 408). Since then, the term scientific literacy has over time been subject to continuous change and diversification of its original meaning (Bybee, 1997, 2002).

In the 1960s, tying the goals of science education to societal ideals, scientific literacy was especially characterized by social aspects of science, such as the recognition of the socio-historical development of science, ethical aspects, cultural dimensions and the social responsibility of science (Bybee, 2002, p. 24). Traditionally, only a small group of students had focused on science education in school, which prepared them for future careers associated with science. In response to this, the National Science Foundation recommended in the 1970s that science education should be rethought with more "emphasis on the understanding of science and technology by those who are not and do not expect to be professional scientists and technologists" (Hurd, 1998, p. 409). Thus, it was along with the 'Science for All' approach in the 1970s (Fensham, 1985; Gräber, 2002; Jenkins, 1999) emphasized that "all students should become scientifically literate if they were to deal effectively as adults with [...] important social concerns" (DeBoer, 1991, p. 174). In this way, the term scientific literacy became "the watchword of the 1970s" and came to be used to describe "an education in science for all youth that was relevant to their lives and that focused on socially important issues" (DeBoer, 1991, p. 174). The understanding of scientific literacy in the 1970s was thus supplemented with aspects such as the nature of science, concepts and processes of science, values, the connection between science and society, interest in science, and scientific skills (Bybee, 2002, 24). However, as the term lacked precision, it was also used to "describe a wide assortment of educational goals" (DeBoer, 1991, p. 174). Shen (1975, pp. 46–49) conceptualized in the 1970s three different, but not mutually exclusive types of scientific literacy, termed as *practical, civic,* and *cultural* scientific literacy. *Practical* scientific literacy refers to the possession of the type of scientific knowledge that can be used to help solve practical problems, whereas *civic* scientific literacy relates to scientific knowledge and understanding that is necessary for informed participation in public life and policy-making. Shen described *cultural* scientific literacy as being motivated by a desire to learn something about science as a major human achievement. In the 1980s, the term scientific literacy was supplemented by aspects referring to scientific and technological processes, research technologies, scientific and technical knowledge, scientific and technical skills and knowledge in personal and social areas, attitudes with respect to science and technology, and the interaction between science, technology and society for technical processes (Bybee, 2002, p. 24). This enhancement came along with the development of science education from an orientation on the disciplines towards a more contextual orientation of science education, opening it up to technological and social issues, which were realized through various approaches of the Science-Technology-Society (STS) movement (Yager, 1993). Later on, a claim for a renewed vision of scientific literacy in view of contemporary issues was formulated (Hurd, 1998, p. 411) and the term scientific literacy has come to refer to a desired general awareness and understanding of science in public (DeBoer, 2000, p. 582).

Based on an emerging emphasis of the relation between science and society, some researchers claim in particular that scientific literacy based science education should focus more strongly on the global and societal dimensions of the applications of science, discussing social issues that contain various points of view, assessments and different options to act (Kolstø, 2001). In this manner, Ramsey (1997, p. 325) states that "citizenship has been and continues to be an ultimate goal of the educational enterprise". Thus, "school should prepare [...] people to exercise the rights and responsibilities of citizenship" as part of an established goal of general education from which science education should not be exempted:

"All students [...] will become citizens [...] [and] will be consumers of the products and services of science and technology. All will assume and be responsible for the benefits and the risks of scientific and technological knowledge, products, systems, and services. All will be decision-makers concerning matters of science and technology, either willfully via participation in democratic decision-making or apathetically via the lack of such participation" (Ramsey, 1997, p. 325).

Today, the need of scientifically literate citizens and the development of scientific literacy for all has been recognized worldwide (UNESCO, 1993), as scientific literacy matters both at national and international levels as humanity faces major challenges such as providing sufficient water and food for all, controlling diseases, generating sufficient energy and adapting to climate change (United Nations Environment Programme, 2012 quoted in OECD, 2013, p. 3). The OECD (2013, p. 3) emphasize that

"many of these issues arise, however, at the local level where individuals may be faced with decisions about practices that affect their own health and food supplies, the appropriate use of materials and new technologies, and decisions about energy use".

As pointed out by the European Commission, the solutions to political and ethical challenges involving science and technology "cannot be the subject of informed debate unless young people possess certain scientific awareness" (EC, 1995, p. 28 quoted in OECD, 2013, p. 3). The European Commission emphasizes that this does not mean "turning everyone into a scientific expert, but enabling them to fulfil an enlightened role in making choices which affect their environment and to understand in broad terms the social implications of debates between experts" (OECD, 2013, p. 28). As knowledge of science contributes significantly to individuals' personal, social, and professional lives, scientific literacy is central to a young person's preparedness for life (OECD, 2013, p. 3).

In recent discourse on desired outcomes of science education for all citizens in the sense of scientific literacy, many voices emphasize socio-scientific facets and see scientific literacy as a premise for social participation and citizenship (Jenkins, 1999; Kolstø, 2001; Sadler & Zeidler, 2009). Also, the applicability of scientific knowledge and the development of competences have come more closely into focus (Gräber & Nentwig, 2002, p. 13). Prevailing conceptions of scientific literacy as a basis for lifelong learning stress

"the development of a general understanding of important concepts and explanatory frameworks of science, of the methods by which science derives evidence to support claims for its knowledge, [...] of the strengths and limitations of science in the real world [...] [and] value[s] the ability to apply this understanding to real situations involving science in which claims need to be assessed and decisions made" (OECD, 1999, p. 59).

Moreover, the OECD underlines both an appreciation of the contribution of science to society and a reflective approach to science (2006, p. 21).

Though the assumption that only few people will achieve this in practice has been a source of criticism of scientific literacy and thus researches do not unanimously agree on its feasibility (Shamos, 1995, 2002), scientific literacy is considered the main purpose of science education within mandatory education and represents a major goal for science education for students worldwide (e.g. Bybee, 1997; DeBoer, 2000; Dillon, 2009; Eckebrecht & Schneeweiß, 2003; Gräber & Bolte, 1997; Gräber & Nentwig, 2002; Koballa et al., 1997; Laugksch, 2000; Millar & Osborne, 1998; NRC, 1996; OECD, 2000). Thus, reforming the curriculum so that students achieve scientific literacy has been and still is a major undertaking (e.g. Hurd, 1998, p. 411, American Association for the Advancement of Science, 1993. p. 323).

However, a review of research literature shows that the term bears many different meanings in the international science educational discourse, and that there are still a variety of attempts to bring normative clarification to the idea of scientific literacy. In a review of the different facets of scientific literacy, Norris and Phillips (2003, p. 225) present a list of references in what ways scientific literacy is most frequently used in literature. Aspects scientific literacy is used for include:

- knowledge of the substantive content of science and the ability to distinguish science from non-science
- understanding science and its applications
- knowledge of what counts as science
- independence in learning science
- ability to think scientifically
- ability to use scientific knowledge in problem solving
- knowledge needed for intelligent participation in science-based social issues
- understanding the nature of science, including its relationships with culture
- appreciation of and comfort with science, including its wonder and curiosity

- knowledge of the risks and benefits of science
- ability to think critically about science and to deal with scientific expertise

Bybee has provided major contributions to the discourse on scientific literacy (1997, 2002). Assuming that scientific literacy consists of different levels, he postulates a hierarchical model. According to this model, every individual can undergo a development along nominal, functional, conceptual and multidimensional scientific literacy, suggesting a continuous progression of the degree of scientific literacy as a lifelong process instead of a dichotomy. Bybee distinguishes four different levels in this competence model. The first level refers to *nominal scientific literacy*. This level includes a certain familiarity with scientific terms and topics. However, this familiarity is still characterized by a lack of adequacy and a naive understanding of theories. The second level is labeled as *functional scientific literacy*. This level corresponds to a stage in which individuals are able to use and recognize scientific terminology and scientific language in an appropriate way, but confined to a particular requirement or activity and without being aware of the function and meaning of the used vocabulary in its further context. As the third level, Bybee determines *conceptual and procedural scientific literacy*. This level refers to relating information to the conceptual ideas which connect the different disciplines and fields of the sciences. Hence, the focus of this level is on the central ideas that characterize the sciences. In addition, this stage also includes a comprehensive understanding of the processes and the nature of science. The fourth and highest level is *multidimensional scientific literacy*. This level of scientific literacy goes beyond vocabulary, concepts and procedural methods and includes further perspectives of the sciences, such as the history of science, the nature of science, and the role of science in personal life and society. Accomplishing the next level of scientific literacy within Bybee's model is represented by achieving the knowledge, skills and insights that correspond to the different stages. Bybee criticizes that throughout time, the different dimensions have been pronounced in an unbalanced way. He underlines that realizing scientific literacy in educational practice should seek an adequate balance between his proposed levels of scientific literacy instead of an overemphasis of one level (Bybee, 1997, pp. 27ff.; Bybee, 2002, pp. 23ff.).

As pointed out by Streller (2009, p. 18), Schenk has criticized Bybee's model (2007, pp. 85–87). She disapproves of Bybee's proposition that through such predefined competence development, the views of the students are mostly ignored, contradictory to the theory of *Bildungsgangdidaktik* (cf. Schenk, 2005). She criticizes that in such an understanding of scientific literacy, the consideration of *Bildung* as a process in which the students not only acquire knowledge but also develop their own personalities and world views through individual progress seems to be missing. However, Bybee's understanding of scientific literacy has received widespread attention in science educational discourse.

At the international level, Bybee's model is used in PISA as a reference for the assessment of scientific literacy in terms of what 15-year-old students should know and be able to do within appropriate personal, social, and global contexts. The concept of scientific literacy applied in PISA is mostly determined by conceptual and procedural aspects and thus corresponds most closely to Bybee's third level labelled as *conceptual and procedural scientific literacy* (OECD, 1999, p. 60). Based on three broad dimensions – scientific knowledge or concepts, scientific processes, and situations or contexts (OECD, 2000, p. 76), PISA 2000 and 2003 defines scientific literacy as

> „the capacity to use scientific knowledge, to identify questions and to draw evidence-based conclusions in order to understand and help make decisions about the natural world and the changes made to it through human activity" (OECD, 1999, p. 60, 2004b, p. 133).

This definition underlines the idea that science education should reach beyond the particular subjects and the mastery of specific curriculum content, leading to more general competences of the learners. With those competences being seen as the basis for the applicability of scientific knowledge in real-life contexts, the main focus of scientific literacy in PISA is "the ability to reflect on and use [...] scientific knowledge, understanding and skills to achieve personal goals and to participate effectively in society" (OECD, 2000, p.10). Moreover, scientific literacy in PISA 2000 includes an understanding of the methods by which science derives evidence to support claims for scientific knowledge, and of the strengths and limitations of science in the real world (OECD, 2000, p. 12). The definition of scientific literacy in PISA 2006 was extended to include affective aspects (OECD, 2006, pp. 21–22) and describes scientific literacy in terms of four interrelated features that involve an individual's

> - "scientific knowledge and use of that knowledge to identify questions, acquire new knowledge, explain scientific phenomena and draw evidence-based conclusions about science-related issues
> - understanding of the characteristic features of science as a form of human knowledge and enquiry
> - awareness of how science and technology shape our material, intellectual, and cultural environments
> - willingness to engage in science-related issues and with the ideas of science, as a reflective citizen" (OECD, 2006, p. 23).

These features of scientific literacy are summarized as

"the capacity of students to identify scientific issues, explain phenomena scientifically and use scientific evidence as they encounter, interpret, solve and make decisions in life situations involving science and technology" as, "in order to participate fully in today's global economy, students need to be able to solve problems for which there are no clear rule-based solutions and also to communicate complex scientific ideas clearly and persuasively" (OECD, 2007b, p. 33).

PISA 2015 further develops the construct of scientific literacy and defines it in terms of three domain-specific competences that a scientifically literate individual would be expected to display. These include the ability to explain phenomena scientifically (e.g. recognizing, offering and evaluating explanations for a range of natural and technological phenomena), to evaluate and design scientific enquiry (e.g. as describing scientific investigations and proposing ways of addressing questions scientifically) and to interpret data and evidence scientifically (e.g. analyzing and evaluating data, claims and arguments in a variety of representations and drawing appropriate scientific conclusions) (OECD, 2013, p. 7). PISA 2015 indicates what knowledge these competences require besides content knowledge ("knowledge of science") by specifying the PISA 2006 notion of "knowledge about science" more clearly, splitting it into procedural and epistemic knowledge (OECD, 2013, p. 10). In this way, scientific literacy in PISA 2015 involves not only knowledge of the major conceptions and theories of science (content knowledge), but also knowledge of the common procedures and methods associated with scientific inquiry (procedural knowledge), and an understanding of the underlying rationale for these procedures and the justification for their use, i.e. the degree to which such knowledge is justified by evidence or theoretical explanations (epistemic knowledge) (OECD, 2013, p. 11). This understanding is summarized as "the ability to engage with science-related issues, and with the ideas of science, as a reflective citizen" (OECD, 2013, p. 7), acknowledging the diverse and multifaceted purposes of science education within compulsory education.

Comparable characteristics are applied by the National Science Teachers Association (1991) or the National Research Council (1996, p. 22):

"Scientific Literacy means that a person can ask, find, or determine answers to questions derived from curiosity about everyday experiences. It means that a person has the ability to describe, explain, and predict natural phenomena. Scientific literacy entails being able to read with understanding articles about science in the popular press and to engage in social conversation about the validity of the conclusions. Scientific literacy implies that a person can identify scientific issues underlying national and local decisions and express positions that are scientifically and technologically informed. A literate citizen should be able to evaluate the quality of scientific information on the basis of its source and the methods used to generate it. Scientific literacy also implies the capacity to pose and evaluate arguments based on evidence and to apply conclusions from such arguments appropriately."

Further elaboration on the term scientific literacy has been carried out by Roberts (2007). He differentiates between two overarching visions of scientific literacy referred to as Vision I and Vision II. Vision I looks at science itself, i.e. its laws, theories and processes, and thus stresses scientific content knowledge. This vision is emphasized e.g. in *Benchmarks for Science Literacy* (AAAS, 1993) and *Science for All Americans* (AAAS, 1990), which attempt to bring operational meaning to scientific literacy. Vision II, on the other hand, refers to situations with a scientific component and focuses on socio-scientific decision making, a perspective of scientific literacy that is also emphasized in PISA (Bybee & McCrae, 2011, p. 8). With contextually embedded issues, Vision II goes beyond the traditional boundaries of science and features in this way a proximity to social, political, economic and ethical perspectives. Thus,

> "[w]hereas articulations of Vision I [...] look to the discipline of science to define what scientifically literate individuals ought to know and be able to do, Vision II [...] looks to situations that present opportunities for individuals to use scientific ideas, processes, and reasoning" (Sadler & Zeidler, 2009, p. 910).

In this way, these visions can at their most extreme been seen as competing interests with respect to the content of the science curriculum (Dillon, 2009, p. 203), covering the spectrum of educating future scientists versus educating future citizens (Bybee & McCrae, 2011, p. 11). However, according to Roberts (2007, p. 11), both visions should be considered for a comprehensive view of scientific literacy in order to achieve a balance between Vision I and Vision II[2].

In the discussions on the term scientific literacy, it becomes apparent that "[s]cientific literacy has become an internationally well-recognized educational slogan, buzzword, catchphrase, and contemporary educational goal" and that there are a variety of ideas subsumed under the concept of scientific literacy (Laugksch, 2000, p. 71). Hence, Millar (2006, p. 1500) concludes that

> "despite the ubiquity of the term "scientific literacy" in current discussions and debates about the science curriculum, its precise meaning is, however, often unclear,

---

2 Several additional discussions of this concept can be found. Miller (1997) addresses in particular the development scientific literacy in the United States. Coll & Taylor (2009) as well as Walberg and Paik (1997) provide overviews of exploring international perspectives of scientific literacy, and Oelkers (1997) elaborates on defining and justifying scientific literacy for everyone. Eisenhart and her colleagues (1996) explore the ways in which scientific literacy has been defined, justified and operationalized in current proposals for science education reform, and Baumert (1997) discusses the term from a German perspective.

and the extent of consensus about the practical implications of adopting it as a central aim of the school science curriculum is uncertain".

Osborne also points out that with "a lack of an explicit and consensually agreed articulation" of scientific literacy (Osborne, 2007, p. 174), differences in meanings and interpretations, as a result may, may have given rise to a view that "scientific literacy is an ill-defined and diffuse concept" (Laugksch, 2000, p. 71). Yet, despite this criticism, scientific literacy appears to underpin the curriculum standards of many countries (Dillon, 2009, p. 201). In addition, Gräber (2002, p.7) argues that despite different approaches, the international discussion on scientific literacy includes a lot of common ideas linked to those of general education in terms of *Allgemeinbildung* and similarly tries to overcome the division of subject-specific contextual knowledge and cross-curricular competences.

In conclusion, scientific literacy as the overall aim of science education can be summarized as

"an evolving combination of the science-related attitudes, skills, and knowledge students need to develop inquiry, problem-solving, and decision-making abilities, to become lifelong learners, and to maintain a sense of wonder about the world around them" (Council of Ministers of Education, Canada, 1997),

representing "a continuum of understanding about the natural and designed world" (Bybee, 1997, p. 63) and the "cognitive capacities for utilizing science/technology information in human affairs and for social and economic progress" (Hurd, 1998, p. 411). In this way, goals of scientific literacy based science education include:

- *to* encourage students at all grade levels to develop a critical sense of wonder and curiosity about scientific and technological endeavors
- to enable students to use science and technology to acquire new knowledge and solve problems, so that they may improve the quality of their own lives and the lives of others
- to prepare students to critically address science-related societal, economic, ethical, and environmental issues
- to provide students with a foundation in science that creates opportunities for them to pursue progressively higher levels of study, that prepares them for science-related occupations, and that engages them in science-related activities appropriate to their interests and abilities
- to develop in students of varying aptitudes and interests a knowledge of the wide variety of careers related to science, technology, and the environment (Council of Ministers of Education, Canada, 1997).

In Germany, aims of science education with respect to enhancing students' scientific literacy are also addressed in the national science education standards in terms of defining different competence areas (KMK, 2005a, 2005b, 2005c). The aims of scientific literacy oriented science education are summarized as making scientific phenomena tangible, understanding scientific language, communicating scientific findings, and dealing with specific methods of scientific inquiry and its limitations, an orientation regarding scientific and technical occupations, and providing the foundations for lifelong and professional learning (KMK, 2005a, p. 6).

In general, DeBoer (1991, p. 240) concludes that science education, as all education, should lead to independent self-activity, empowering individuals to think and to act, and providing new ideas, investigative skills that contribute to self-regulation, personal satisfaction, and social responsibility, interconnected knowledge, intellectual skills allowing individuals to work with what is known, and an awareness of the contexts within which that knowledge and those skills apply.

2.1.4 General Science and the Tradition of Subject Differentiation

In all European countries, science education is to varying degrees compulsory throughout primary and secondary education, beginning as one general integrated subject area and taught in this way throughout the entire period of primary education[3]. It intends to foster children's curiosity about their environment by providing them with basic knowledge about the world and by giving them the tools with which they can investigate it (EC, 2011, p. 60). In several countries, this approach is continued for one or two years in lower secondary education, but is split into separate subjects towards the end of lower secondary education[4]. In this way, lower levels of education mostly feature integrated science teaching. Some countries even teach science as one integrated subject all the way through compulsory school[5]. However, in lower secondary education, science teaching is usually split into separate subjects such as biology, chemistry, and physics (EC,

---

3 Exceptions are Denmark and Finland, where a division into individual subjects begins before the end of primary education (EC, 2011, p. 61).

4 In Belgium (German-speaking part), Bulgaria, Estonia, Spain, France, Malta, Slovenia and Liechtenstein, science teaching is continued as an integrated program at lower secondary education level, but towards the end of this level, science education also takes place through the different science subjects (EC, 2011, p. 61).

5 In six European education systems (Belgium – French and Flemish Communities, Italy, Luxembourg, Iceland, and Norway) science is taught as an integrated subject entirely throughout both primary and lower secondary education levels (EC, 2011, p. 61).

2011, p. 61), which represent the most commonly taught science subjects in European countries (EC, 2011, p. 7). Yet, while dividing science teaching into different subjects, several countries underline the interrelation between biology, chemistry, and physics in order to encourage teachers to teach in interdisciplinary contexts as often as possible (EC, 2011, p. 62). Many countries apply mixed forms of these approaches, particularly for lower secondary education (EC, 2011, p. 61). Thus, on the one hand, there are education systems mostly determined by biology, chemistry, and physics as separate subjects, on the other hand, many education systems feature curricula with an integrated science subject, so that integrated and subject-specific approaches are present in the European countries to varying extents (EC, 2004, p. vii). Beyond Europe, approaches of integrated science education are practiced in countries of the English-speaking world (NRC, 1996; Prenzel, 2010, p. 21). Countries with integrated science education at the lower secondary level usually offer an overarching science subject, bringing together in its curriculum themes and areas of the different science disciplines. Integrated science education is generally subsumed in subjects such as "science", but labels often include references to "environment and technology" (EC, 2011, p. 62), "physical, life and earth science", "science as inquiry", "science in personal and social perspectives", "history and nature of science", and „processes of science" (Bybee, 2002, pp. 23–24), or are represented by approaches such as "Science–Technology–Society (STS)" (Yager, 1993). Examples of other terms describing the variety of curricular arrangements and degrees of integration of science in the context of general education include interdisciplinary, transdiciplinary, multidisciplinary and cross-curricular science education (EC, 2011, p. 59; A. Kremer & Stäudel, 1997, pp. 56–57; Labudde, Heitzmann, Heiniger, & Widmer, 2005, p. 103). Following the European Commission (2011, p. 59), the term integrated science education is applied in this work to represent all curriculum arrangements that merge elements from a minimum of two science disciplines.

Many science education researchers trace back the establishment of general science as a subject in school to the general science movement in the United States at the beginning of the 20th century (DeBoer, 1991, p. 87). As pointed out by DeBoer (1991, p. 88),

"the notion that an education should be well rounded and focused on the single aim of character development led to a number of attempts to create unifying themes to relate two or more areas of the curriculum".

In this course, general science as a school subject was established with a focus on the interests and the intellectual development of the young adolescents and with respect to science in the context of everyday life (DeBoer, 1991, p. 88).

Later on, during the curriculum movement and the education reforms of the 1950s and 1960s in the United States, which aimed at enhancing student achievement to achieve a larger workforce in science and science related professions, the focus of science education in the Anglo-American world shifted, however, to following a structure-of-the-discipline approach, giving students in science education a sense of the fundamental ideas of a discipline as early as possible (DeBoer, 1991, p. 159), dealing with the nature of scientific research, basic concepts and encouraging students "to think and act like scientists within the structure that was established (DeBoer, 1991, p. 171). Yet, especially since the 1960s, models of general science that focused on the environment, concepts, and processes, received a renaissance in the Anglo-American world (Häußler, 1973, p. 48). On an international level, the development of science education mostly proceeded from an orientation on the disciplines towards context orientation (Gräber, 2002, p. 3).

In Germany, there is a long tradition of subject differentiation of science education, particularly in secondary education (e.g. Riquarts et al., 1994; Riquarts & Wadewitz, 2001). The establishment of different science subjects in school education in Germany dates back to educational reforms during the 19th century, in the course of which instruction in the natural sciences known as *Realienunterricht* became a regular part of the curriculum (Schöler, 1970, p. 156). Its curriculum was mostly based on the science disciplines (Schöler, 1970, p. 150). Throughout the last decades, science education in Germany has been organized in more flexible ways. The debate on the introduction of integrated science education in Germany and the construction of integrated science curricula started as part of the educational reform in 1970 with Anglo-American models of general science entering the German curriculum discussion on science education. First, interdisciplinary approaches and recommendations for integrated science education received massive resistance of the science education research community and of defenders of traditional science education practice who saw integrated science education as a threat to the autonomy of the traditional science subjects biology, chemistry and physics. In this way, the integrated science subject could not be established in the German school system despite the development of teaching materials and concepts and was for many years only realized on side tracks such as project weeks and in the comprehensive school specific subject "science". Only at the end of the 1980s, in the course of an attempt to redefine education with respect to *Allgemeinbildung*, reform initiatives with integrative character regained attention, and ideas were specified and more specifically realized in practice. Both old and new approaches to the integration of science subjects inspired the didactic discussion about a renewal of science education in

Germany in many ways, so that intensified reform initiatives and pilot projects in this field took place (Frey, 1989; A. Kremer & Stäudel, 1997, pp. 52–54). The Institute for Science Education at the University of Kiel (Institut für Pädagogik der Naturwissenschaften [IPN]) with the *Integrated Science Curriculum* project played an important role in the development of the idea of integrated science education in Germany (Häußler, 1973). This project included the development of basic structures for the design and the implementation of integrated science curricula and made new impacts on the attempt to overcome subject-inherent structural rigidities in science education. As PISA and TIMSS have given new impetus to the discussion of science education in Germany as well, the idea of an integrated science curriculum was taken up and included in the framework curricula by the curriculum committees of several federal states. However, there is still an ongoing debate on the organization of science education in the German science education research community as well as on a European level, both in terms of integration versus subject differentiation and on the question whether scientific literacy is fostered better within the frame of discipline-oriented subjects such as biology, chemistry, and physics, or within an integrated science subject (EC, 2011, p. 59; Labudde & Möller, 2012, p. 15; Prenzel, 2010, p. 21).

The common instructional approach in science education of differentiating according to subjects obtains its legitimacy and its structural organization from disciplinary core areas of specialized contents and their specific sequence to provide systematic and comprehensive access to scientific concepts (Ramsey, 1997, p. 399). Usually, discipline-based science curricula are organized on the basis of the knowledge, methods, structure, and language of the academic disciplines (e.g. AAAS, 2001; KMK, 2005a, 2005b, 2005c). The MNU[6] stresses that although the scientific method forms the common basis of biology, chemistry and physics, each of these disciplines contains its own scientific perspective. These distinct perspectives are expressed on the one hand in subject-specific concepts and terms. On the other hand, despite their similarities with respect to the nature of their world view and the way of inquiry, the three disciplines emphasize different methodical aspects. While physics is characterized by a high degree of mathematical formalization and abstraction, chemistry is distinguished through methods such as working and observing on the particle level and thinking in sub-microscopic models using its own, symbol-oriented scientific language. For biology, characteristic features are the detection of the history of life and the theory of natural selection as an explanatory principle. Subject-specific

---

6 Deutscher Verein zur Förderung des mathematischen und naturwissenschaftlichen Unterrichts (German Association for the Promotion of Mathematics and Science Education)

differences are thus inherent in the subject-specific concepts, methods and explanations. Hence, the three subjects do not look at the world through the same "lenses" but consider different strands of the scientific interpretation of the world (MNU, 2003, p. 10). In this way, certain reservations are expressed with respect to realizing approaches of integrated science education.

General concerns towards integrated science education focus on the lack of empirical evidence for the positive impact of this approach on the students' motivation and their performance. Some researchers argue that students might gain a less basic and conceptual understanding in integrated science education, as certain discipline-specific topics are covered in a less detailed way or not at all. Another critical aspect when considering integrated approaches are the teachers' skills and content knowledge, as teachers are usually educated in a limited number of academic disciplines and might feel uncomfortable with the integration of disciplines into their teaching for which they are not trained or qualified (EC, 2011, p. 60). Further objections refer to the difficulty of achieving discipline-specific goals within integrated science, the complexity of interdisciplinary themes and resulting difficulties in their consideration, the preference of biological topics, and the risk of a reduction of the number of science lessons (Labudde & Möller, 2012, p. 15).

On the other hand, it is often criticized that school science is usually fragmented into different strictly isolated disciplines, and it is claimed that science education is in this way failing to provide students with a coherent picture (Christidou, 2011, p. 146). More specifically, it is argued that a division into disciplines is too compartmentalized and hence counterproductive to serve the general education needs of students (AAAS, 2001, p. 86). Also, some researchers claim that despite carrying their own special features, the science disciplines share significant epistemological, conceptual and methodological similarities with respect to scientific thinking and working (Prenzel, 2010, p. 22). Thus, these scholars claim that scientific literacy based science education needs to transcend the familiar perspective of discipline-based school science (Deng, 2007, p. 136).

From a European perspective, two lines of argument are proposed. First, integration in science education seems to correspond to a "common sense", since knowledge and experience are not divided into different compartments in real life either. This claim emphasizes that the traditional discipline boundaries do not meet contemporary requirements and that scientific research itself is becoming increasingly integrated and intertwined so that education in different science disciplines cannot be undertaken well in isolation from the others. The second line of argument focuses on the process of knowledge construction. The teaching of science in a holistic approach and making connections between the different disciplines is considered as a process leading to new ways of thinking that links various abilities, enhances critical thinking and allows for a deeper understanding of scientific issues (EC, 2011, p. 59). Further frequently expressed arguments include the assumption that students' pre-concepts are not formed within discipline structures, possibilities of project-oriented science learning, the promotion of transdisciplinary competences, and the development of students' willingness to engage – in line with Klafki's education theory (see 2.1.1) – with epoch-typical key problems of humanity (Labudde & Möller, 2012, p. 15). These arguments with respect to interdisciplinary approaches of science education are often related to scientific literacy (e.g. Bauer, 1992, p. 113; Gräber, 2002, p. 7).

In Germany, general thoughts supporting integrated science education are expressed in similar ways. It is claimed that furthering scientific literacy within mandatory school education does not require an exact representation of the science disciplines in school or a transition of a simplified version of different science disciplines into school, as this is a misinterpretation of the relationship between academic disciplines and school subjects (Klafki, 1995b, p. 194). Rather, enhancing scientific literacy in school demands an intensive consideration of the real-life contexts of the students, consulting science with respect to its potential and limitations for solving real-life issues (Frey, 1974, p. 16; A. Kremer & Stäudel, 1997, p. 57). Therefore, it is crucial to temporarily neglect traditional structures of the subjects and reveal new relations and linkages instead. In this manner, it is emphasized that a strict division of science education into different science subjects does not reflect reality and is thus an artificial construct that would lead to narrow-mindedness and isolated considerations of scientific issues (A. Kremer & Stäudel, 1997, p. 57). Schaefer (2010a, p. 11) moreover argues that when considering general abilities with respect to *Allgemeinbildung*, only certain parts of science related abilities are subject-specific, whereas a significant part of such abilities is of more overarching nature.

One example of realizing integrated science education in Germany within general education which gives new impetus to the underlying ideas of integrated science education in Germany is the PING project (*Praxis integrierter naturwissenschaftlicher Grundbildung – Practising Integration in Science Education*) (Bünder, 1997; Lauterbach, 1992b, 1993). The PING project was developed in 1989 at the Institute for Science Education at the University of Kiel (Institut für Pädagogik der Naturwissenschaften [IPN]) and is an integrated science curriculum for the development of scientific literacy within general education for grades 5-10. It aims at a balance between systematic science learning within subject-differentiation and real-life-oriented approaches, bringing together seemingly distinct realms and addressing problem-based themes from the different science subjects in an interdisciplinary way through action-oriented approaches of science teaching (Bünder, 1997, pp. 399–400; Lauterbach, 1993, pp. 244–246), following the tradition of Anglo-American integrated science approaches of the 1970s (A. Kremer & Stäudel, 1997, p. 55). PING is based on the assumption that science education takes place in the reflected relation of the individual to nature, other people, culture, and to him or herself. The overall basis of PING is the relationship between the individual and nature. In particular, this includes aspects such as methods of recognizing, treating, and interacting with nature and technology, the comparison of different views, applications of common principles and concepts, and the coordination of discipline-based terms. PING addresses these aspects through age-appropriate themes that are developed with respect to the real-life experiences of the students (Lauterbach, 1993, pp. 244–247). Through its design and conceptual structure, PING also reflects and contributes to the basic ideas of an education for all in the sense of Klafki's understanding of *Bildung* (Lauterbach, 1993, pp. 244ff.).

What generally distinguishes an integrated science curriculum from subject differentiated science teaching is that other elements than the disciplines determine how the content will be organized – elements such as natural phenomena, social and environmental issues, or other cross-cutting themes. However, this does not mean that discipline-based curricula necessarily neglect certain issues or that integrated curricula disregard knowledge and methods from the disciplines. Principally, the same set of specific educational goals could be pursued within either form of organization (AAAS, 2001, p. 87).

Often, controversies about integrated versus subject based science are based on an oversimplified comparison between a school science subject and the corresponding disciplines (Prenzel, 2010, pp. 22–23). In fact, the science disciplines in school are constituted within a reference frame of content and problems coordinated, matched and oriented towards educational goals – not necessarily reflecting the structure of the corresponding science disciplines, which are limited to pursuing scientific goals. Prenzel further points out that in schools of general education, the traditional science subjects indeed provide a necessary requirement for systematic knowledge acquisition regarding the current state of science, but this premise is not sufficient for the universally demanded general education, which schools of compulsory education are obliged to. Hence, Prenzel concludes that adjusting, limiting, and reflecting on learning contents, and also transcending subject boundaries should always take place with respect to the targeted educational goals. It is crucial that students become primarily familiar with the basic issues, concepts, terms and theoretical and methodological approaches of science and understand the special features of each disciplinary perspective. The characteristic features of the scientific disciplines can be made obvious by including an integrative perspective, and considering phenomena and problems from different perspectives. For that reason, an integrated perspective on science should include the disciplinary approaches and perspectives and ensure the development of several competences at the same time. In this way, Prenzel points out that in the case of subject-differentiation, the science subjects in school still have to fulfill a mutual supply function, which requires at least curricular coordination and linkages between the science subjects.

2.1.5 Enhancing Science Education in Europe – The PROFILES Project

On the basis of overtly expressed concerns that science education in school is widely perceived as irrelevant and difficult, the European Commission funds projects to increase "students' interest and attainments levels while at the same time stimulating teacher motivation" in the European Commission's 7th Framework Program (EC, 2007, p. 2). This general framework encompasses education-oriented investigations and supports actions that aim at improving the level of scientific literacy in young people.

PROFILES ("Professional Reflection-Oriented Focus on Inquiry-based Learning and Education through Science") is one of several inquiry-based science projects funded by the 7th Framework Program of the European Commission in the field of science and society. The consortium of the PROFILES Project involves 22 partner institutions from 21 different European or Europe-associated countries, including Freie Universität Berlin (Germany), University of Tartu (Estonia), Weizmann Institute of Science (Israel), Universität Klagenfurt (Austria), Cyprus University of Technology (Cyprus), Masaryk University (Czech Republic), University of Eastern Finland (Finland), University College Cork (Ireland), Universita Politecnica delle Marche (Italy), University of Latvia (Latvia), Maria Curie-Sklodowska University (Poland), University of Porto (Portugal), Valahia University of Targoviste (Romania), University of Ljubljan (Slovenia), University of Valladolid (Spain), Fachhochschule Nordwestschweiz (Switzerland), Dokuz Eylül University (Turkey), Universität Bremen (Germany), ICASE[7] (UK), Karlstad University (Sweden), University of Copenhagen (Denmark), and Ilia State University (Georgia). The project is coordinated by the Department of Chemistry Education at Freie Universität Berlin (Bolte, Streller, et al., 2012; PROFILES, 2010b).

The overall focus of PROFILES is "the promotion and dissemination of IBSE[8] through innovative learning environments and (long-term) teacher training courses to raise self-efficacy of the participating science teachers to take owner-

---

7 International Council of Associations for Science Education
8 Inquiry Based Science Education

ship of more effective ways of teaching to foster students gains – supported by stakeholders" (PROFILES, 2010b). On this basis, PROFILES aims at disseminating a modern understanding of scientific literacy and encouraging new approaches to the practice of science teaching, following an education-through-science approach. Central ideas of science education in the education-through-science approach of PROFILES include an orientation towards the needs and interests of the students, following an IBSE-approach and strong emphasis on scientific inquiry. It takes up themes from real-live contexts, addressing age-appropriately central concepts of science, subject-transcending themes, taking into account the competence areas of the national education standards in a balanced way, guiding students to reflected and appropriate judgment, and aiming at intrinsically motivated learning beyond science lessons (PROFILES, 2010b).

To attain these aims, the project is composed of eight interrelated work packages that address the different aspects of the project. With work package 3 (WP3), which focuses on stakeholder involvement and interaction, PROFILES gives emphasis to examining the views of different stakeholders from science education related areas, asking for their perspectives on what aspects of science education they consider as relevant for the scientific literate individual as a member of society. The significance and relevance of investigating such views is underlined by the fact that the theme of science and society represents a central pillar in the seventh Framework Program by the European Commission. The context of PROFILES allows for collecting the views and opinions of such stakeholders from 21 countries (Bolte et al., 2011; Bolte, Streller, et al., 2012). The stakeholders' views are collected by means of the Delphi method, a tool of group discussion which provides a systematic approach to involving a wide range of stakeholders and bringing together views and opinions from various perspectives in a collective decision making process (see 2.3.3). Within work package 3, this endeavor is carried out in the context of the International PRO-FILES Curricular Delphi Study on Science Education, which involves in the PROFILES partners' various stakeholders in their countries who are particularly affected by curricular aspects of science education. Such stakeholders are represented by students, science teachers (including science education students at university, trainee science teachers, science teachers and trainee science teacher educators), science education researchers, scientists, and actors in education policy and administration.

The aim of the International PROFILES Curricular Delphi Study on Science Education is to engage these stakeholders in reflecting on contexts, contents and aims of science education and to identify in this way aspects of modern science education that are considered relevant, meaningful, and desirable for the scientifically literate individual of today and in the near future. By applying the Delphi method, this study systematically collects and investigates in three consecutive rounds the views and opinions of different stakeholders. Planned and carried out in the context of PROFILES, the International PROFILES Curricular Delphi Study on Science Education consists of several national curricular Delphi studies being conducted independently from each other by the PROFILES consortium partners in their countries. This provides the opportunity not only to investigate the stakeholders' views on a national level, as done within this study, the Berlin Curricular Delphi Study in Science (Bolte & Schulte, 2014a, 2014b; Schulte & Bolte, 2012, 2013a, 2013b). Furthermore, the framework of PROFILES also allows for identifying similarities and differences between the opinions of stakeholders from different countries (Bolte & Schulte, 2014a, 2014b, 2014c; Gauckler, Bolte, & Schulte, in press; Gauckler, Schulte, & Bolte, 2014; Schulte, Bolte, et al., 2014; Schulte & Bolte, 2012, 2013a, 2014b; Schulte, Georgiu, Kyza, & Bolte, 2014), holding out prospects for insights into a common European perspective on science education as well. Within PROFILES, taking into account these different stakeholders' views and opinions also provides the opportunity to enhance the development of learning and teaching materials and the preparation of teachers' CPD[9] programs as well as to bridge the gap between the science education research community, science teachers and other local actors, allowing for a stronger cooperation between the different stakeholder groups. In this manner, the stakeholders in PROFILES – involved through the International PROFILES Curricular Delphi Study on Science Education – can be seen as partners in the development, evaluation and dissemination of the projects activities and outcomes as well (Bolte et al., 2011; Bolte, Streller, et al., 2012).

2.1.6 Summary

As pointed out at the beginning of this chapter, a need for science education was expressed along with the emergence of science as a definable style of inquiry (Oliver et al., 2001, p. 4). During the 19th century, science was thus increasingly established as part of the curricula of schools of general education (DeBoer, 2000, p. 583; Schöler, 1970, p. 9). On the basis of recognizing the contribution of science to education in general (cf. von Engelhardt, 2010), this development took place in the course of a growing notion of the practical importance of science in a world becoming increasingly dominated by science. This development was triggered by recognizing the importance of the inductive process of observing the natural world, drawing conclusions from it, carrying out independent

---

9 Continuous Professional Development

inquiries and investigations in the laboratory, developing an attitude of independence from arbitrary authority, and participating more fully and effectively in an open democratic society (DeBoer, 2000, p. 583).

Today, science is a fundamental and indispensable dimension of education in general, as science education is recognized to contribute to the individual's ability to actively partake in society and opinion-making with respect to science related issues. The central place of science in the school curriculum of mandatory education is also justified by the enormous impacts of science and scientific ideas on our everyday lives and culture. Therefore, in most European countries, science education takes place within the frame of general and common shared educational goals, such as developing the students, both individually and socially, and developing competences, knowledge, skills and attitudes that are deemed to be important for future citizens by each country (EC, 2004, p. vii).

The scientific perspective as an essential component of *Allgemeinbildung* is captured in the concept of scientific literacy, which is considered the main and overarching aim of science education. In German, scientific literacy is translated as *naturwissenschaftliche Grundbildung* and can be defined as a combination of overarching and science-related attitudes and competences.

Common features and aims of a science education with an emphasis on scientific literacy include the process of interaction with and reflection of the world, discovering throughout this interaction one's own role in society, culture and nature, being prepared to act in reflective and responsible ways, and being able to actively partake in social discourse and opinion-making. It is also argued that "the development of positive attitudes and the stimulation of curiosity [...] are as important as the development of the conceptual understanding of the subjects" (EC, 2004, viii). Chemistry, biology, and physics represent predominant areas of school science as a regular part of the curriculum. In contrast, with respect to enhancing students' scientific literacy, integrated science education underlines a perspective on science as a comprehensive and interdisciplinary endeavor.

An example of efforts to improve science education and to promote students' scientific literacy through an education-through-science approach is represented by the European PROFILES project.

2.2 Science Education and Curriculum

In the first part of this chapter, I will introduce the concept of curriculum. It will be defined with respect to the purpose of this study. Furthermore, I will discuss the concept of curriculum theory with reference to the German tradition of *Didaktik*. Following these considerations, I will describe different curricular dimensions of science education.

2.2.1 The Concept of Curriculum

Due to the essential role of instruction and teaching in education, curricular matters are central topics of concern in society (Pinar, 2009, p.33). In Germany, the topic of curriculum particularly entered the discussion with reference to scientific literacy after the outcomes of international comparisons such as PISA (e.g. OECD, 2000, 2004a) and TIMSS (e.g. Baumert et al., 2000) and the subsequent debate on the poor performance of the German students.

In general, the term curriculum covers a broad range of meanings, referring to a general sketch of what should happen in schools and to the conditions of learning as well as denoting the students' actual day-by-day experiences. Moreover,

"papers in the journals speak of a "planned" curriculum, which is different from the 'taught' curriculum (the instruction actually delivered to students), which differs from the 'learned' curriculum (what students actually learn)".

Overall, curriculum is thus often understood in all of its meanings – as it is planned by teachers and administrators, delivered by the teachers, and experienced by the students, as an overview of the scope and sequence of student experiences, and as a detailed delineation of learning experiences (AAAS, 1993, p. 318).

The roots of curriculum theory can be traced back to Comenius (Schaller, 1995, p. 57), who is said to have established the first curriculum of modernity that became a guideline for the following centuries (Gundem, 1995, p. 46). In 1952, Weniger (1952 quoted in Westbury, 1995, p. 241) described the task of the curriculum as "to establish the goals of education, and to select and concentrate what used to be called instructional material". In this way, the term curriculum was originally used to generally refer to the progress of educational measures as well as to learning objectives (Frey, 1970, p. 15). In Anglo-American terminology, curriculum relates to aims, measures, forms of organization and objects that appear in the context of teaching (cf. Posner, 1995). Yet, curriculum is not to be understood synonymously with the teaching process (Frey, 1970, p. 14).

When addressing the term curriculum, two different traditions of approaching curricular matters have to be mentioned – the German tradition of *Didaktik* and Anglo-American curriculum theory (Hopmann & Riquarts, 1995b, p. 9). In Germany, curricular thought has been closely linked to the tradition of *Didaktik*. American curriculum theory only entered the academic discussion in Germany in the 1960s (Hopmann & Riquarts, 1995b, p. 21). Taking account of the complexity that has taken place with respect to the development of these traditions (Hopmann & Riquarts, 1995b, p. 11), Hopmann and Riquarts (1995b, p. 9) state that "both attitudes exist in such a variety of forms that neither can easily be reduced to a single pattern". Yet, in spite of this variety, they argue that each side operates with certain common presuppositions about the relationship between curriculum and instruction. Differences between these approaches refer to several aspects at the level of lesson planning as well as at the level of research and theory (Hopmann & Riquarts, 1995b, p. 26; Westbury, 1995, p. 233), with *Didaktik* being primarily concerned with instruction itself, while curriculum mainly covers aspects such as study plan, school books, instruction concepts etc. (Hopmann & Riquarts, 1995b, p. 21).

However, other scholars in the field of *Didaktik* and curriculum research do not share the view that there are substantial differences between *Didaktik* and curriculum or curriculum theory in Anglo-American usage. For example, Klafki (1995, p. 187) points out that a comparison between attempts to define *Didaktik* and corresponding attempts to define curriculum theory reveals large overlaps, as they are concerned with a parallel set of issues[10].

With the German curriculum discussion of the 1970s and 1980s not systematically distinguishing between *Didaktik* and curriculum research (Hopmann & Riquarts, 1995b, p. 20), Frey describes the term curriculum as the representation of education over a certain period of time in terms of a consistent system with multiple areas for purposes of planning and optimal realization of instruction (1970, p. 15). In addition, Frey (1980b, pp. 21–22) distinguishes between two approaches to the concept of curriculum; a restricted term and a comprehensive term. The restricted term of curriculum essentially means the preparation of teaching, a syllabus or another offer of means that can lead to an educational process. In this understanding, a curriculum is a disposition for a planned educational process and includes all conditions that lead to this process. Thus, the curricular process begins with the first decision or establishment of conditions for an educational process. How the process should be designed is described in this approach, for example, by naming learning goals and objectives. The second, more comprehensive term extends the meaning of the restricted definition by including the realization of the curriculum and its impact. In this way, the term curriculum is also compared to a social process. In addition, the more comprehensive term includes further aspects with respect to the methodology of curriculum development. Hence, Frey (1980a, p. 23, translation following Hopmann & Riquarts, 1995b, p. 20) describes curriculum as the answer to

---

10 As it is beyond the aim of this chapter to trace the dialogue between Didaktik and curriculum theory in a more comprehensive scope, the discussion on differences is not elaborated on in more detail. Further insights into the dialog between Didaktik and curriculum theory are provided by Hopmann and Riquarts (1995a).

"how can learning situations be developed, implemented and evaluated, which in the horizon of their societal and objective environment and of the individual self-interpretation of the learners are justified and at the same time optimally guarantee the self-development of all concerned [...] before, during and after the envisaged learning process?".

A central element in this approach is curricular justification. According to the work of Frey (1980b, p. 23), justification in a curricular context refers to developing a claim of validity, which is constituted both by theoretical reflections and through involving appropriate stakeholders (see 2.3.2).

Following Frey (1980b, p. 22), this study employs the term curriculum both in its restricted and comprehensive sense. Hence, one product of this study could be seen as a framework for science education providing aspects that the scientifically literate individual should deal with within his or her process of education. Thus, this study leads to science curriculum related features that could serve as a basis and orientation for further realization of scientific literacy based science education. On the other hand, this study also provides insights with regard to curriculum development in the sense of the comprehensive understanding of the concept of curriculum.

2.2.2 Curriculum Elements of Science Education

It is widely recognized that within the context of education, the issue of curriculum content is a fundamental one (Phillips & Siegel, 2013). This notion relates to an argument by Klafki (2000, p. 90), who claims that despite the complexity of the concept of *Bildung*, the implicit questions dominating the reflections on the content of *Bildung* remain the same:

"What objectifications of human history seem best suited to open a person who is engaged in his or her own *Bildung* [...] to the possibilities and duties of an existence in humanity?"

With respect to the relationship between the process of *Bildung* and curriculum, educational content should also be determined at the level of frameworks such as guidelines and curricula besides the level of decisions in instruction of individual schools and teachers (Klafki, 1995b, p. 194). Hence, the question must be asked what attitudes, insights, knowledge, and abilities young people require in order to be able to develop their capacity for self-determination, co-determination and solidarity in their present and future lives. In this way, curriculum design involves a variety of aspects that need to be considered.

As "education for scientific literacy is not a new goal, but one that has come into prominence globally through major curriculum initiatives" (Ratcliffe &

Millar, 2009, p. 946), a number of authors "have attempted to clarify the curricular orientation and instructional emphasis of scientific literacy as a purpose of science education" (Bybee, McCrae, & Laurie, 2009, p. 866). Thus, several curricular movements, designed to promote scientific literacy, have emerged in the field of science education. These include, for example, the Science-Technology-Society movement and related efforts promoted under slogans such as public understanding of science, humanistic science education, context-based science education, and socio-scientific issues, continuing what is described as progressive science education (Sadler & Zeidler, 2009, p. 910). In *Designs for Science Literacy* (AAAS, 2001), scientific literacy is described as a curricular vision that is "used to convey the normative, ideological basis for determining the subject matter of school science" (Deng, 2007, p. 135) and which has "significant implications for the choice of curriculum content and the way it is structured" (Ratcliffe & Millar, 2009, p. 946).

However, with respect to the question what in particular should constitute the subject matter of the school curriculum for scientific literacy, it is indicated that although definitions of scientific literacy abound in literature and scientific literacy has widely been accepted as a central goal of school science education in the 21st century (cf. 1.3.1), there is no consensus about what the curricular implications of scientific literacy are (Deng, 2007, p. 134). In fact, with scientific literacy being characterized in terms of attributes such as knowledge, skills, and dispositions of a scientifically literate person, or types of literacy involved, there are various definitions capturing the rich meanings of scientific literacy as an educational goal. Nonetheless, they convey very little about the meaning of scientific literacy as an educational goal as it is translated into curricular structures and into classroom practice. Hence, the meaning of scientific literacy should be investigated more closely from the curriculum perspective as well (Deng, 2007, p. 134). According to Ratcliffe and Millar (2009, p. 946), there is at least some agreement that scientific literacy involves three identifiable strands when it comes to considering scientific literacy as a curriculum aim: science concepts and ideas, processes of scientific enquiry, and the role of science in the social context.

Following the approach of Häußler et al. (1980), curricular dimensions of desirable science education can be specified within contexts and situations, topics and concepts, and competences and attitudes. This classification is based on the assumption that generally, all life situations, appearances and actions related to science can be educationally relevant (Häußler et al., 1980, p. 50). On a broader level, this approach can be related to the work of Robinsohn (1975, p. 45), who defines the task of curriculum research as the identification of skills that serve to cope with life situations and the identification of educational content that supports the enhancement of such qualifications. Similar classifications of curricular dimensions of science education can also be derived from the PISA 2006 definition of scientific literacy (OECD, 2006, p. 25), which is characterized as consisting of four interrelated components representing different scientific contexts, contents, competences, and attitudes. The classification of these components is based on the question "[w]hat is it important for citizens to know, value, and be able to do in situations involving science and technology?" (OECD, 2006, p. 20).

As the overarching aim of this study is to identify aspects of science education that are considered desirable and meaningful for students to achieve scientific literacy, the following chapters will discuss contexts and situations, topics and concepts, and competences and attitudes as specifications of curricular aspects of science education.

2.2.2.1 Contexts and Situations

There is growing recognition of the significance of contexts in science education (e.g. Elster, 2007). The notion of contexts as the starting point for the development of a scientific understanding in science education dates back to the early 1980s (Bennett & Lubben, 2006, p. 999; Fensham, 2009, p. 884) and is closely related to achieving overall aims of science education (Waddington, 2005, p. 306), as contexts also allow for addressing aspects of scientific literacy that are not subject-specific (Häußler et al., 1980, p. 50). In particular, embedding science teaching in contexts is seen as a potential way of improving student motivation and interest (EC, 2011, p. 64; Millar, 2005, p. 325) and as effective in enhancing students' scientific literacy (e.g. Council of Ministers of Education, Canada, 1997; Deng, 2007, p. 136). The Canadian *Common Framework of Science Learning Outcomes* (Council of Ministers of Education, Canada, 1997) emphasizes that the development of scientific literacy is supported by meaningful instructional contexts that engage students in active inquiry, problem solving and decision making, as it is through such contexts that students discover the significance of science in their lives and come to appreciate the interrelated nature of science, technology, society, and environment. In this way, contexts are essential to student learning (Finkelstein, 2005, quoted in Gilbert, 2006, p. 970).

Originating from the Latin verb "contextere" ("to weave together"), the related noun "contextus" expresses "coherence", "connection", and/or "relationship". Thus, the general function of "context" is to describe circumstances that provide meaning (Gilbert, 2006, p. 960). According to Duranti and Goodwin (1992, pp. 6–8, quoted in Gilbert, 2006, p. 961), an educational context can be described to feature four attributes. These relate to the framework of a certain event or general phenomenon (setting), actions and measures (behavioral environment), terminology and language framework (language), and background (knowledge).

As context can have several meanings (Whitelegg & Parry, 1999, p. 68), in many cases the term situation is used instead or in addition to context (e.g. Baumert et al., 1999, p. 4). Following Kortland (2011, p. 5), the particular purpose of contexts in science education is

> "to embed science knowledge in a collection of practical situations [...] showing, first of all, that science relates to everyday life and enables us to understand practical applications and socio-scientific issues, and, secondly, that science content has a personal and/or social relevance in enabling thoughtful decision making about everyday life behavior".

Thus, contexts, situations – and motives as a related term – can be retrieved from a variety of areas such as nature, leisure, politics, science, consumption, household, public, world view, and culture (Hoffmann & Rost, 1980, p. 65). Typically, they include social, economic, environmental, technological and industrial applications of science and are selected on the basis of their relevance to students' everyday life (Bennett, Gräsel, Parchmann, & Waddington, 2005, p. 1523).

Although the selection of particular contexts may vary, it is recommended that the overall scope and focus of instructional contexts in science education should include three broad areas of emphasis to provide potential starting points for engaging in an area of study (Council of Ministers of Education, Canada, 1997):

-   scientific inquiry, in which students address questions about the nature of things, involving broad exploration as well as focused investigations
-   problem-solving, in which students seek answers to practical problems requiring the application of their science knowledge in new ways
-   decision-making, in which students identify questions or issues and pursue science knowledge that will inform the question or issue

In Europe, it is also suggested that science education processes should be embedded in contexts. Usually, this involves a relation to social and real-life contexts or situations that students are likely to encounter as citizens, such as contemporary societal problems, environmental concerns, the application of scientific achievements, and knowledge to everyday life as starting points for the development of scientific ideas (EC, 2011, p. 9). In this way, recommend contextual issues that teachers should address in science education at primary and lower secondary level in several European countries include, for example, "science and the environment/sustainability", "science and everyday technology", "science

and the human body", "science and ethics", "embedding science into its so-
cial/cultural context", "history of science", and "philosophy of science" (Eurydice, 2011, p. 66). As

> "[m]odern definitions of [...] scientific literacy similarly emphasize the importance
> of recognizing and understanding the contexts in which [...] science operate[s] and
> the forces that shape these fields of human activity" (OECD, 2000, p. 15),

situations and contexts represent the third aspect of scientific literacy in the PISA
2000 framework, besides processes and concepts (Baumert et al., 1999, p. 4).
Real-life related situations in PISA 2000 are classified into problems that can
affect people as individuals (personal level, such as food and energy consumption), as members of a local community (public level, such as drinking water
treatment or search for power plant sites) or as citizens of the world (global level,
such as global warming, loss of biodiversity). Additionally, situations within the
framework of PISA 2000 relate to the advancement of scientific knowledge and
the influence of this knowledge to societal decisions (historical relevance)
(Baumert et al., 1999, p. 8). PISA identifies "science in life and health", "science in earth and environment", and "science in technology" as three main context areas relevant for scientific literacy (OECD, 2006, pp.12-25, see also Bybee,
Fensham, & Laurie, 2009; Bybee & McCrae, 2011; Drechsel, Carstensen, &
Prenzel, 2011). These contexts are specified within five sets of life situations:
"health", "natural resources", "environment", "hazards", and "frontiers of science and technology", which are, as in PISA 2000, further divided into personal,
social or global settings (OECD, 2006, pp. 26–27). Fensham (2009, p. 893) underlines that such contexts are rarely mono-disciplinary, but provide starting
points for both single subject and integrated science teaching. Sadler (2004, p.
523) describes contexts that

> "encourage personal connections between students and the issues discussed, explicit-
> ly address the value of justifying claims and expose the importance of attending to
> contradictory opinions"

as most valuable contexts in terms of starting points for learning processes of and
about science.

In Germany, projects such as biology in context (cf. Bayrhuber et al., 2007),
chemistry in context (cf. King, 2012; Parchmann et al., 2006) and physics in
context (cf. Duit & Mikelskis-Seifert, 2010) provide particular contributions on
the discussion about how meaningful science education can succeed by embedding scientific concepts, methods and knowledge in contexts that support the
students in developing and enhancing their competences.

2.2.2.2   Concepts and Topics

The validity of the justification regarding the inclusion of particular content in educational is a fundamental issue with respect to curriculum development (Phillips & Siegel, 2013). Hence, several attempts have been made to address the issue of relevant teaching and learning content for enhancing students' scientific literacy (e.g. AAAS, 1990, 1993, 2001). PISA 2000 defines scientific concepts from different content areas as fundamental with respect to scientific literacy (OECD, 2000, p. 75). According to the PISA approach to scientific literacy, concepts in science education enable students to give meaning to new experiences as they relate them with what they already know and support students to understand aspects of the natural and man-made world. PISA determines the selection of scientific concepts according to their relevance for everyday life situations, sustainable meaning in the future of the students as citizens, and allowing for linkages to scientific processes (Baumert et al., 1999, pp. 5–6). Topics from physics, chemistry, biology, earth and space science, and technology complying with such criteria are represented in PISA by knowledge of science categories (classified according to physical systems, living systems, earth and space systems and technology) and knowledge about science categories (classified according to scientific inquiry and scientific explanations). They include:

"Physical systems
- structure of matter (e.g. particle model, bonds)
- properties of matter (e.g. changes of state, thermal and electrical conductivity)
- chemical changes of matter (e.g. reactions, energy transfer, acids/bases)
- motions and forces (e.g. velocity, friction)
- energy and its transformation (e.g. conservation, dissipation, chemical reactions)
- interactions of energy and matter (e.g. light and radio waves, sound and seismic waves)
Living systems
- cells (e.g. structures and function, DNA, plant and animal)
- humans (e.g. health, nutrition, subsystems [i.e. digestion, respiration, circulation, excretion, and their relationship], disease, reproduction)
- populations (e.g. species, evolution, biodiversity, genetic variation)
- ecosystems (e.g. food chains, matter and energy flow)

- biosphere (e.g. ecosystem services, sustainability)

Earth and space systems

- structures of the Earth systems (e.g. lithosphere, atmosphere, hydrosphere)
- energy in the Earth systems (e.g. sources, global climate)
- change in Earth systems (e.g. plate tectonics, geochemical cycles, constructive/destructive forces)
- earth's history (e.g. fossils, origin and evolution)
- earth in space (e.g. gravity, solar systems)

Technology systems

- role of science-based technology (e.g. solve problems, help humans meet needs and wants, design and conduct investigations)
- relation between science and technology (e.g. technologies contributing to scientific advancement)
- concepts (e.g. optimization, trade-offs, cost, risk, benefit)
- important principles (e.g. criteria, constraints, innovation, invention, problem solving)

Scientific inquiry

- origin (e.g. curiosity, scientific questions)
- purpose (e.g. to produce evidence that helps answer scientific questions, current ideas/models/theories guide enquiries)
- experiments (e.g. different questions suggest different scientific investigations, design)
- data type (e.g. quantitative [measurements], qualitative [observations])
- measurement (e.g. inherent uncertainty, replicability, variation, accuracy/precision in equipment and procedures)
- characteristics of results (e.g. empirical, tentative, testable, falsifiable, self-correcting)

Scientific explanations

- types (e.g. hypothesis, theory, model, law)
- formation (e.g. data representation, role of extant knowledge and new evidence, creativity and imagination, logic)
- rules (e.g. must be logically consistent; based on evidence, historical and current knowledge)
- outcomes (e.g. produce new knowledge, new methods, new technologies; lead to new questions and investigations)" (OECD, 2006, pp. 32–33).

However, scientific concepts are formulated on many different levels, ranging from comprehensive descriptions of science up to long lists of generalizations, as they are often presented in descriptions of curricular requirements (Baumert et al., 1999, p. 5).

2.2.2.3 Competences and Attitudes

A third realm of curricular dimensions includes all levels of competences and attitudes, representing cognitive, affective and operational components (Spada, 1980, p. 113). Frequently, the term "scientific habits of mind" is used for describing processes associated with the application of scientific, mathematical and technological knowledge to everyday life in terms of competences and attitudes, as they relate directly to a person's outlook on knowledge and learning and their ways of thinking and acting (AAAS, 1993, p. 322).

On a general level, the competence is understood as "the comprehensive precondition for problem-solving in a specific field of reality" (Schaefer, 2010b, p. 16). From a curricular perspective, Klieme et al. (2007, p. 21) point out that competences reflect basic demands on students in a certain domain. Referring to abilities in the context of general education, Klafki (2007, p. 63) defines basic competences of an educated person in the sense of *Allgemeinbildung* as argumentation skills, criticism and self-criticism, empathy, and complex thinking.

In light of the different and sometimes vague use of the term competence[11] and related terms such as meta-competence or key competencies both in everyday language and in research literature, substantial conceptual clarification is provided by Weinert (2001b). As competence covers – from innate personality traits to an acquired extensive body of knowledge, and from cross-curricular key qualifications to subject-specific skills – a wide range of denotations (cf. Weinert, 2001a), Weinert recommends a pragmatic approach in which competences should be conceptualized as the necessary prerequisites for meeting complex demands. More specifically, he describes competence as available or learnable cognitive skills and abilities of an individual to solve problems, and associated motivational, volitional and social skills and abilities required to successfully and responsibly apply problem solving in a variety of situations (Weinert, 2001b, pp. 27–28). Thus, competences cannot be reduced to their cognitive components, but include as integral parts ethical, social, emotional, motivational, attitudinal and behavioral facets, which together allow for effective action in specific situations (Rychen, 2008, p. 16). In this way, a competence can be seen

---

11 Due to the problem of clear distinctions, related terms such as competences, skills, abilities, and qualifications are used in this work interchangeably.

as a disposition that enables people to successfully solve certain types of problems, that is, to cope with situations of specific requirements. According to Weinert, the individual shapes of competence are determined by different interrelated facets such as ability, knowledge, understanding, action, experience, and motivation (2001b, pp. 27–8). The internal structure of a competence is defined by the demands, tasks, activities, interrelated attitudes, values, knowledge and skills that make effective action possible (Rychen & Salganik, 2002, p. 5).

The OECD DeSeCo[12] project provides an interdisciplinary approach to theoretically grounded conceptual foundations for identifying competences needed for individuals to lead responsible and successful lives in a modern democratic society and for society to face the challenges of the present and the future (cf. OECD, 2001, p. 3; Salganik, Rychen, Moser, & Konstant, 1999, p. 6). In line with Weinert's definition, the DeSeCo defines competences for applications such as in PISA (2001b, pp. 27–28) in a functional, external and demand-oriented approach, describing a competence as "the ability to meet individual or social demands successfully or to carry out an activity or task" that is "developed through action and interaction in formal and informal educational contexts" (OECD, 2002, pp. 8–9). Moreover, DeSeCo underlines the personal and social requirements individuals face and emphasizes that a competence is acquired and developed throughout life (OECD, 2005, p. 17).

In the context of school education, competences are primarily subject-based. This means that both subject-specific and more general, interdisciplinary competences can only be developed on the basis of and within the framework of certain subjects. However, the different subjects also have to be aware of their significance and have to fulfil their responsibility with respect to the development of central and universal competences. In this sense, the MNU argues that science education should promote both subject-specific and subject-transcending competences (MNU, 2003, p. 4). Describing scientific literacy as being composed of a domain-specific set of competences, Gräber et al. (2002, p. 137) distinguish between three dimensions of competences: knowledge (language and epistemological competence), acting (learning, communication, social, and procedural competence) and assessment (ethical and aesthetic competence). PISA 2000 (OECD, 1999, p. 62) applies a view of scientific literacy that enables it to be defined in terms of a number of scientific competences closely connected to conceptual content:

---

12 Definition and Selection of Competencies

- recognizing scientifically investigable questions
- identifying evidence need in a scientific investigation
- drawing or evaluating conclusions
- communicating valid conclusions
- demonstrating understanding of scientific concepts

In PISA 2006, these competences are modified and extended with a more process oriented focus (OECD, 2007b, p. 37):

- identifying scientific issues (recognizing issues that are possible to investigate scientifically, identifying keywords to search for scientific information, and recognizing the key features of a scientific investigation)
- explaining phenomena scientifically (applying knowledge of science in a given situation, describing or interpreting phenomena scientifically and predicting changes, as well as identifying appropriate descriptions, explanations, and predictions)
- using scientific evidence (interpreting scientific evidence and making and communicating conclusions, identifying the assumptions, evidence and reasoning behind conclusions, and reflecting on the societal implications of science and technological developments).

In a similar way, prerequisites for developing scientific literacy are described in terms of skills required for scientific and technological inquiry, solving problems, communicating scientific ideas and results, working collaboratively, and making informed decisions (Council of Ministers of Education, Canada, 1997).

Prominent core competences in these lists are thinking, communicating, problem solving, and inquiry. The statements of values and purposes accompanying these lists refer to connectedness, resilience, achievement, creativity, integrity, responsibility, and equity. It is, however, criticized that

> "for science educators and science teachers, this language of the knowledge society and what it means for education is almost entirely foreign" and "[e]ven problem solving is not elaborated in discipline-specific terms, but in generic strategies of various types" (Fensham, 2007, p. 116).

Fensham continues to point out that although PISA science has made a great contribution by defining competences that can be developed in science education, it is, as part of evaluative research that describes and measures wanted outcomes, not a curriculum. Nonetheless, PISA science can provide essential starting points for curriculum design (2007, p. 117).

In Germany, the national science education standards also describe the contribution of science education to students' scientific literacy through certain competences that students should achieve (KMK, 2005a, 2005b, 2005c). For each of the three common science subjects (biology, chemistry, physics), it distinguishes four areas of key competences: content knowledge, inquiry, communication, and evaluation. The domain content knowledge is represented by subject-specific basic concepts. Within this domain, it describes which phenomena, facts, terminology, applications, and laws the students should know and be able to relate to each other with respect to the different concepts. The domain scientific inquiry refers to methods and experimental techniques within scientific observation and the application of models which students should be able to carry out. The domain communication addresses aspects of task-oriented and issue-related accessing and sharing of information. The domain evaluation refers to recognizing and evaluating scientific issues in different contexts.

Following Weinert (2001b, pp. 27–28), competences include ethical, social, emotional, motivational, and behavioral components. Consequently, attitudes are closely associated to competences. In the tripartite view shared by Rosenberg and Hovland (1960), attitudes consist of cognitive, affective, and behavioral components. In this way, they constitute the affective basis of education (Schaefer, 2010b, p. 16). Attitudes play a significant role in students' scientific literacy because they

> "represent a complex system of cognitions, feelings, and inclinations towards action, and [...] influence students' continuing interest in and dispositions for positive and constructive responses towards science and science-related issues" (Bybee & McCrae, 2011, p. 23),

both in general and with respect to issues that affect them in particular. Sjøberg and Schreiner (2010, p. 4) underline that affective dimensions of science education are also relevant from the perspective of life-long learning and society. These considerations provide the rationale for including attitudes as a major part of curricular dimensions of science education.

As part of science competences and one of the four interrelated components forming the assessment of scientific literacy, a person's disposition towards science in terms of attitudes as beliefs, motivational orientations, self-efficacy, and values is classified in PISA (OECD, 2007b, p. 39) as an individual's

- support for scientific enquiry (acknowledging the importance of considering different scientific perspectives and arguments, supporting the use of factual information and rational explanations, expressing the need for logical and careful processes in drawing conclusions)

- self-belief as science learners (handling scientific tasks effectively, overcoming difficulties to solve scientific problems, demonstrating strong scientific abilities)
- interest in science (indicating curiosity in science and science-related issues and endeavors, demonstrating willingness to acquire additional scientific knowledge and skills, using a variety of resources and methods, demonstrating willingness to seek information and have an ongoing interest in science, including consideration of science-related careers)
- responsibility towards natural resources and environments (showing a sense of personal responsibility for maintaining a sustainable environment, demonstrating awareness of the environmental consequences of individual actions, demonstrating willingness to take action to maintain natural resources).

With respect to the citizenship dimension of scientific literacy, the *Canadian Framework of Science Learning Outcomes* in particular emphasizes that students should be encouraged to develop attitudes supporting the responsible acquisition and application of scientific knowledge to the mutual benefit of self, society, and the environment (Council of Ministers of Education, Canada, 1997).

In Germany, the MNU (2003, p. 10) classifies attitudes relevant in the context of (science) education as required by ability to participate in scientific and social life within three domains:

- learning and thinking
    - emotional: joy, curiosity, interest, perseverance, sensitivity, self-criticism
    - cognitive: various forms of thought such as analytic, synthetic, systemic, typological, causal, convergent, divergent, exclusive, inclusive etc.
- attitudes to oneself and to the environment:
    - communication, cooperation, tolerance
    - responsibilities to oneself and others
    - protection and care for one's own health and that of others as well as the preservation of the environment
- attitudes towards science and technology:
    - use, opportunities, sustainability
    - risks, safety
    - responsibility

With respect to the affective basis of science education, Bybee (2008, p. 570) moreover emphasizes that one important goal of science education is the development of students' attitudes supporting "their attending to scientific issues and the subsequent acquisition and application of scientific and technological knowledge to personal, social, and global benefit".

2.2.3 Summary

The issue that lies at heart of curricular research relates to the question how educational situations that are legitimated within the perspective of their social context and the individual development of the learners can be developed, realized and evaluated. Hence, the central aim of curriculum research is to find and apply methods through which contexts, contents, and abilities necessary for coping with different life situations can be determined. Frey points out the significance of curriculum development and criteria for the process of curriculum construction and development. Three areas can be distinguished as relevant curricular domains with respect to science education. The first domain refers to contexts in which processes of science education can be embedded and situations in which scientific knowledge and abilities are relevant. Scientific or science related topics and concepts represent the second domain. The third domain includes competences and attitudes with respect to scientific literacy, consisting of both cognitive and non-cognitive dimensions.

As society and particularly education systems and education policy are required to increasingly promote skills that enable people to deal responsibly and competently with different challenges, the question what skills are considered essential for future democratic societies is of growing importance (Rychen, 2008, p. 15). The OECD states that defining and selecting valuable, useful and legitimate competences is ultimately the result of a process in which researchers are partners among other stakeholders such as policy makers, practitioners, and representatives of the economic and social world (OECD, 2001, p. 4).

2.3 Curricular Processes in Science Education

Several elements of curriculum processes need to be considered with respect to curricular processes in science education. In order to take a closer look at determining factors for curricular processes in the context of science education, I will in the following sections illustrate elements that classify a curriculum process and the corresponding stages. Moreover, I will define central stakeholders in a science-related curriculum processes. In addition, I will introduce and describe the Delphi technique as a suitable instrument of a curricular process related to science education.

2.3.1   Components of a Curriculum Process

Tyler (1971) distinguishes three curriculum determinants with respect to considering curriculum content as relevant for educational purposes. These include individual relevance, social relevance, and scientific relevance. However, these determinants are characterized by a high degree of generalization, which is not sufficient to serve as the only legitimizing criteria (J. Mayer, 1992, p. 17). According to Frey (1980b, pp. 24–27), a curriculum process is constituted by three components, referring to focus and objectives, interaction and discourse, and reflection. Relating these components to curricular processes in science education, Frey (1980b, p. 24) points out that, first, principles related to general education as established in society need to be considered. These include, for example, the development of the individual and society as a major focus of education. Also, education should contribute to taking on responsibility in greater social contexts, guided by overarching ideals such as solidarity and equality. These objectives also apply to science education, through which – as part of general education – the realization of general and overarching educational goals can be pursued (cf. 1.2). Yet, a curricular process is not about an indoctrination of such objectives. Instead, through appropriate measures, it should be made possible that such objectives are considered and discussed within a curricular process (Frey, 1980b, p. 25). Following Häußler and his colleagues (1980), in the curricular approach of this study this notion is accounted for through the overall method, that is, through an open initial question, the selection of participants, and the iterative way of processing the questions throughout several rounds (cf. 6.3).

With regard to the second component, interaction and discourse, it can be argued that the experiences and insights of experts in a certain domain represent a valuable knowledge basis from which important information for a given issue can be obtained (Häder, 2009, pp. 92–94). Hence, with respect to the basic curricular question regarding what aspects of science education are appropriate for both the learners' personal development, it is important that as well other persons than those being professionally involved with science are included in a curricular process towards desirable science education. It is thus necessary to involve different stakeholders with a range of expertise and backgrounds (see 2.3.2) who relate to the aforementioned objectives of science education (Frey, 1980b, p. 25). Additionally, the interaction of such different stakeholders is an important aspect as it is also essential that curricular discussions of desirable science education develop and provide the opportunity to be addressed in a discursive way (cf. Frey, 1980b, p. 26).

Furthermore, a curricular process requires specific reflection, the third component. In the course of such processes, the element of reflection features a constructive effect in dealing with the interaction within the frame of the objectives. This demand avoids that desirable aims and contents of science education are determined through consultation of the total population, as democracy alone cannot guarantee a legitimate educational concept, no more than a reproduction of discipline-specific expertise can (cf. Frey, 1980b, p. 26). In this way, the systematic involvement of selected groups of participants should be enhanced by diverse interaction and reflection. Relevant stakeholders for a curricular process concerning the question of desirable science education are discussed in the following.

2.3.2 Central Stakeholders of Curriculum Processes in Science Education

Emphasizing scientific literacy as the overarching goal of science education rather than discipline-based purposes poses the question of whose voice should be heard with respect to the science curriculum during the compulsory years of schooling (Symington & Tytler, 2004, p. 1404). Generally, there are several groups with a stake in science education. Such stakeholders, being directly or indirectly affected by science education, often have multiple demands as they deal with issues of science education from different perspectives (Roberts, 1988, p. 28). Hence, science education is a field with different stakeholders having both an interest in and an influence on the curricular matters of science education (AAAS, 1993, 2001; e.g. Aikenhead, 2003; Häußler et al., 1980; Sjøberg & Schreiner, 2010).

According to Aikenhead (2003, pp. 15ff.), groups that decide or should decide what is relevant regarding educational perspectives in science curricula include academic scientists and education officials ("wish-they-knew science"), people facing real-life decisions related to science and curriculum policy researchers ("need-to-know science"), people with careers in science-based industries and professions ("functional science"), mass media and internet ("enticed-to-know science"), experts interacting with the general public on real-life matters pertaining to science, such as public health experts, environmental experts, and economics ("have-cause-to-know science"), students at school ("personal curiosity science"), and people engaged in the field of culture (science-as-culture").

With respect to the field of science communication, Burns et al. (2003, p. 184) distinguish in a similar way six relevant, partly overlapping groups in society. These include "scientists" (industry, academic community and government), "mediators" (science communicators, journalists and other members of the media, educators, and opinion-makers), "decision-makers" (policy makers and scientific and learning institutions), the "general public" (the three groups above, and other interest groups such as students in school), the "attentive public" (the part of the general community already interested in – and reasonably well-informed about – science and scientific activities), and the "interested public" (composed of people who are interested in but not necessarily well-informed about science and technology). Some researchers have criticized the failure to involve relevant members of society in curricular processes in science, as

> "from the literature it is apparent that little is known concerning the views of the community generally about the purposes that science education should set out to achieve" (Symington & Tytler, 2004, p. 1404).

Given the multiple purposes of science education as identified with respect to scientific literacy (cf. 2.1.3), it can be argued that in a debate about desirable aspects of science education, it would be appropriate to seek the opinions of further members of society affected by science, capturing in this way "a community voice that would represent responsible and informed views" (Symington & Tytler, 2004, p. 1405). This claim to extend the scope of stakeholders given a voice in science curricular discussion has received widespread support by a number of scholars and shows a growing awareness of the important role of different members of society in developing appropriate science curricula for scientific literacy (Symington & Tytler, 2004, p. 1404).

However, with defining the public as every person in society, it has to be acknowledged that *the public* with its needs, socio-economic settings and cultural conditions is a very heterogeneous group. Also, it has to be recognized that not every person in society is affected by or involved in science education. Hence, the question arises which members of the public are directly or indirectly affected by and involved in science education and can thus be considered as relevant stakeholders. Representing a within-school view, students and teachers with science subjects surely should have a significant say about what happens in schools in relation to science learning (Symington & Tytler, 2004, p. 1404). In this way, it is argued that a comprehensive and integrated recognition of the voices of students and teachers is necessary to make informed, research-based decisions on designing school science curricula and teaching (Christidou, 2011, p. 141). This view is supported by taking into account the process of education as "an open, complex and recursive system of an intersubjective nature, within which relations and communication between members of the classroom community (i.e. teacher and students) create new, common worlds and contribute to the co-construction of meaning and the constitution of students' identities" (Christidou, 2011, p. 142). Thus, students and science teachers can be identified as two important stakeholder groups with respect to science education. However, while acknowledging the importance of the views of students and teachers, it needs to be recognized that these two groups cannot alone be expected to represent a broader community perspective in the context of science education (Symington & Tytler, 2004, p. 1404). Therefore, additional members of the public affected by and involved in issues of science education and the school science curriculum, such as scientists, the science education research community, and education policy, have to be consulted (Roberts, 1988, p. 27). The following sections describe these groups and the rationale for considering them as relevant stakeholders for science education in more detail.

2.3.2.1 Students

Students are obviously important stakeholders in the field of science education (e.g. Beattie, 2012; Fielding, 2001, 2004; Jenkins, 2005, 2006; Klafki, 2007; Meyer & Reinartz, 1998; Roberts, 1988; Ruddock et al., 2003; Rudduck & Fielding, 2006), as

> "there is something fundamentally amiss about building an entire [education] system without consulting at any point those it is ostensibly designed to serve" (Cook-Sather, 2002, p. 3 quoted in Jenkins, 2006, p. 1).

However, in most societies, aspects that are both important and salient within a given domain, such as science education, are usually defined by the academic community (Osborne et al., 2003, p. 693), which suggests that the voices of students represent "a crucial element still too often overlooked" (Nixon, Martin, McKeon, & Ranson, 1996, p. 270 quoted in Jenkins, 2006, p. 1) or even entirely absent from science curriculum related debates (Osborne & Collins, 2001, p. 442). Yet, with respect to young people's contributions to today's and tomorrow's world, there is a growing recognition that students are capable of insightful and constructive analysis of their experiences of learning in school and are able to make contributions to the development of strategies for improving learning and raising achievement (Ruddock et al., 2003). Thus, scholars increasingly notice that students have a right to be heard and have something worthwhile to say about their school experiences including their science education (Ruddock, 2003, p. 1).

On the basis of these considerations and with students as the main and final users in each educational system, it seems reasonable to argue that curricular matters of science education must also be approached from the perspective of the learner and should satisfy the needs of these learners rather than only the science education community or adult society.

The assumptions about the important role of student voice can also be linked to the German theory of *Bildungsgangdidaktik* (Meyer & Reinartz, 1998; Schenk, 2005), which especially takes into consideration the individual and his or her process of *Bildung*. With the focus on the individual in light of the theory of *Bildungsgangdidaktik*, Meyer (2005, p. 18) defines *Bildung* as a social process in the course of which the individual develops. In the context of formal education, this understanding of *Bildung* emphasizes the students' perspectives in the process of education and also underlines the importance of exchange and communication between adolescents and adults within this process. From a German *Bildungs*-oriented perspective, Klafki (1995b, p. 194) also emphasizes the necessity to mediate between the current interests and experiences of the learners, their current problems within their everyday lives on the one hand and the perspectives beyond this context with respect to the young peoples' future tasks and opportunities, both as individuals and as part of a larger society on the other hand. However, Klafki stresses that student-orientation alone is as misplaced as the establishment of instruction solely from the perspective of an educating generation claiming to be able to anticipate what the next generation will need in the sense of attitudes, insights, abilities and skills to cope with their future. On the basis of these arguments, it becomes apparent that both students and "adult" stakeholder groups represent relevant voices in science education.

2.3.2.2   Teachers

One of the central findings of research in teacher education and expertise is that teachers are central predictors for the realization of effective and modern education and that they play an essential role in implementing innovations (EC, 2007; Hattie, 2009, 2012). In science education, teachers are key figures in the formation and reorganization of students' conceptions and attitudes towards science, as teachers' views and beliefs determine to a large extent their teaching practices (Christidou, 2011, p. 146). The development of research and teaching material as well as policy recommendations are further areas of teachers' involvement and driving force. Sjøberg and Schreiner (2010, p. 2), as well as Roberts (1988, p. 27), point out that this particularly applies to those teachers who are organized and involved in teacher networks and science teacher associations or other networks such as science societies and science centers, which can have a further impact, especially through projects, journals, conferences, or formulating and promoting position statements. In light of these notions, the importance of teachers as stakeholders and decision makers in science education and the central role they play in the development of scientifically literate citizens is widely acknowledged (AAAS, 1997; Millar, 1996; Roberts, 1988; Schreiner & Sjøberg, 2004).

2.3.2.3 Science Education Researchers

Further stakeholders in science education are represented by researchers in the academic field of science education as a professionalized field with academic degrees and positions, research centers, professional associations at regional, national, and international levels, a high number of international professional conferences, and also several national as well as international academic journals (Sjøberg & Schreiner, 2010, p.2).

Science education researchers – who may also be science teachers and/or working in teacher training – can be considered to be evidence providers of factors that affect science education through their research studies, as they can be expected to hold informed views in their area of research. Also, science education researchers play a significant role in handling the education of pre-service teachers at university. Recognizing that their science education students are likely to be teaching for more than 30 years into the future, science education researchers can be said to have the most forward looking perspectives among the different stakeholder groups in science education (PROFILES, 2010a, p. 20). Closely related to the field of science education research are authors and contributors of science textbooks and curriculum actions, as they are involved in producing materials for student and teacher use (Roberts, 1988, p. 27).

2.3.2.4 Scientists

An additionally relevant stakeholder group in the field of science education are scientists (PROFILES, 2010a, p. 21; Sjøberg & Schreiner, 2010, p. 2). This group includes research scientists such as university scholars and staff members (Roberts, 1988, p. 27), as they influence the teaching of science and textbooks in terms of so-called fundamental ideas (PROFILES, 2010a, p. 21). Moreover, the group of scientists refers to professionals in science-related industry and workplaces, who might as managers, decision makers or employers hold particular interest in the enhancement of scientific literacy (PROFILES, 2010a, p. 21; Sjøberg & Schreiner, 2010, p. 2). In this way, this group is often referred to as the "science community" or as "science practitioners" (Burns et al., 2003, p. 184). The professional bodies of scientists are organized in numerous associations on regional, national, or international levels. They provide several policy documents and position statements related to school science education, as their associations and academies often contain sub-groups dealing with school science and science in the public (Sjøberg & Schreiner, 2010, p. 3). Of all stakeholders, scientists working within the science field in the public sector or in industry have the most practical viewpoint from a "life skills" perspective, as they deal with science-related attributes that are important in the workplace (PROFILES, 2010a, p. 21). However, as these stakeholders do not have a direct impact on those who usually act as decision makers in curricular matters of science education, their voice is often ignored (PROFILES, 2010a, p. 21).

2.3.2.5 Education Policy and Administration

Policy makers and curriculum developers are an additionally important stakeholder group, as they determine both guidelines and boundaries for the teachers through the intentions of the curriculum, external examination goals, regulations, or the setting of standards for science teaching in schools. Thus, their views have great impact on the practice of science education (PROFILES, 2010a, p. 21). Stakeholders for science education in the group of education policy includes persons in Ministries of Education with a direct remit towards science teaching, science curriculum developers, curriculum committees in school systems appointed by government departments, further formal national educational authorities and education administration, or others such as local town councils with an interest in the enhancement of scientific literacy (PROFILES, 2010a, p. 21; Roberts, 1988, p. 27).

2.3.3 The Delphi Technique as an Instrument for the Investigation of Curricular Aspects of Science Education

A process involving the investigation of curriculum content in terms of desirable science education as targeted in this study represents a complex issue. With respect to the central elements of a curriculum process (cf. 3.1), this issue can be explored by obtaining opinions of experts from different areas in a systematic way, as following Frey (1980a, p. 30) such a process requires on the one hand involving a number of stakeholders through an appropriate way and on the other hand accounting for the requirements of systematic reflection and participation in the different stages of this process. A suitable method for such a complex endeavor emerges through the Delphi technique, which is considered an appropriate approach to gain access to the knowledge of different experts in a given domain (Häder, 2009; Häder & Häder, 2000). Studies suggest that in terms of accuracy, the Delphi method as one of various group-decision making processes outperforms unstructured group judgment as well as standard interacting groups (Bolger, Stranieri, Wright, & Yearwood, 2011; Rowe & Wright, 2001) and is thus frequently applied to a range of judgment problems (Ayton, Ferrell, & Stewart, 1999). Having already proven its applicability in the field of education and curriculum-related research (Bolte, 2003a, 2003b, 2008; Edgren, 2006;

Farmer, 1995; Häußler et al., 1980; Heimlich, Carlson, & Storksdieck, 2011; Judd, 1972; Marshall, Currey, Aitken, & Elliott, 2007; J. Mayer, 1992; Osborne et al., 2003; Reeves & Jauch, 1978; Rice, 2009; Robertson, Line, Jones, & Thomas, 2000; Rockwell, Furgason, & Marx, 2000; So & Bonk, 2010; van Zolingen & Klaassen, 2003; Welzel et al., 1998; Wicklein, 1993; Yang, 2000), the Delphi method is considered a suitable tool for collecting and structuring experts' opinions on educational issues (Ammon, 2009, p. 462; Häder, 2009, p. 238; Häder & Häder, 2000, p. 14; Linstone & Turoff, 1975a, p. 10).

2.3.3.1   The Delphi Method

The Delphi technique is a special form of written consultation and represents a highly structured method of group communication. More specifically, it is a widely acknowledged way of accessing, collecting, organizing and condensing views and opinions of a panel of experts on a given topic throughout several consecutive rounds in a systematic way (Aichholzer, 2002; Ammon, 2009; Ayton et al., 1999; Becker, 1974; B.B. Brown, Cochran, & Dalkey, 1969; Bernice B. Brown, 1968; Dalkey, 1969; Häder, 2009; Häder & Häder, 2000; Kenis, 1995; Linstone & Turoff, 1975c; Murry & Hammons, 1995; Seeger, 1979). Considering the variety of Delphi applications, the Delphi method is defined in a number of different ways, according to the particular context and depending on which aspect is emphasized (Ammon, 2009, p. 459). Linstone and Turoff (1975a, p. 9) have defined the Delphi technique as

> "a method for structuring a group communication process so that the process is effective in allowing a group of individuals, as a whole, to deal with a complex problem".

This elicitation of expert opinion is a central feature of the Delphi method and allows for an aggregation of available knowledge, experience and judgment of experts (Häder, 2009, p. 21). Overall, the results of Delphi studies thus serve to gain insights about aspects that are difficult to determine and that can in this way provide guidance and support for the accomplishment of tasks and the realization of goals. For these reasons, the Delphi method is in particular applied in view of complex decisions in order to gain access to the expertise of persons with different perspectives on a given issue (Ammon, 2009, p. 461).

The name of the Delphi method derives from the antique oracle of Delphi in Greece, which provided predictions through which political and social influence was exercised during the 5th and 6th century A.D. (Häder, 2009, p. 13). The first use of an approach with reference to the Delphi oracle in modern times relates to the context of forecasting results of dog or horse races in 1948 (Häder, 2009, p.

15). In the United States, the Delphi method was developed as a tool for group communication concerning the access of expert opinion for technological forecasting by the Air Force-sponsored RAND Corporation[13] in the context of defense research (Dalkey & Helmer, 1963, p. 458; Linstone & Turoff, 1975a, p. 16). The aim of this first Delphi study was to

> "obtain the most reliable consensus of opinion of a group of experts [...] by a series of intensive questionnaires interspersed with controlled opinion feedback" (Linstone & Turoff, 1975a, p. 16).

The initial rationale for the Delphi method is based on Dalkey's assumption referred to as the 1+n argument, which suggests that 1+n persons can provide at least as much information as one person, but most likely, 1+n minds would be able to offer more information (Häder, 2009, p. 38). The first published Delphi study appeared in 1964 in the *Report on a Long Range Forecasting Study* by the RAND Corporation (Häder, 2009, p. 15). In the following years, the Delphi method was adapted in a variety of areas and became a widely used technique both in the public and in the private sector. Much of the subsequent use of the Delphi method refers to the purpose of predicting technological developments. Additional common areas of application include several other fields with judgment issues such as industrial planning, telecommunications, education, tourism, economic development, medical progress and health policy, regional development, environmental research, and government (Häder & Häder, 2000, pp. 13–15; Linstone & Turoff, 1975a, p. 10).

With various modifications and the great diversification of methodical features especially since the 1970s (Häder & Häder, 2000, p. 15), several types and sub-types of Delphi emerged (Häder, 2009, p. 19). Häder (2009, pp. 29–35) distinguishes between four different types of Delphi studies with respect to its general objectives. The first refers to the generation and aggregation of ideas as the main focus, representing an all-qualitative approach. The second type relates to the goal of achieving predictions that are as accurate as possible of an unclear issue or uncertain situation. The third type applies to Delphi studies that primarily aim to empirically identify and display a group of experts' views on a diffuse issue. The results of such studies serve, for example, to gain specific conclusions for carrying out interventions to react to an issue identified in this way. The fourth type gives particular emphasis to inducing group communication processes through the feedback component that lead to establishing consensus among the participating experts. The type applied within this study refers to the third

---

13 The RAND (Research and Development) Corporation is a think tank in the United States providing expertise on economic, social and defense issues (The RAND Corporation, 2013).

type, as the purpose of this study mainly focuses on empirically identifying and displaying a group of experts' views on a complex issue. However, while there are many varieties of Delphi, common to all are design considerations including aspects such as the initial question, expertise criteria, sample size, questionnaire design, iteration, number of rounds, anonymity of the participants, feedback, structure of information flow, panel mortality, and mode of interaction (see 2.3.3.1.).

In general, a Delphi study undergoes four distinct phases. The first phase is characterized by an exploration of the subject with the aim of determining the objectives of the study and reaching an operationalization of the overarching question. The second phase refers to the process of establishing a questionnaire for collecting expert opinions in order to gain an understanding of how they view the issue. The third phase involves an evaluation of the previous results by the central monitoring team of the study and feedback of the results to the experts. This phase includes possibly repeating the collection of the experts' views, look- ing at their potential change of view on the basis of the feedback of the previous results until a predefined ending criterion is reached. The last phase takes place when all previously gathered information has been analyzed and the evaluations have been fed back for consideration, and concludes with a final evaluation (Häder, 2009, p. 25; Linstone & Turoff, 1975a, pp. 5–6). The general procedure of a Delphi study is addressed in more detail in 3.3.1.10.

The "classical" Delphi approach can be characterized by several key fea- tures (Häder, 2009, p. 25; Häder & Häder, 2000, p. 17):

- a formalized question format as a means for the collective opinion-making process taking place within the study
- involvement of individuals that can be considered experts within the field of investigation
- a fixed group of participants who are consulted with respect to a certain issue
- anonymity of the participants
- statistical aggregation of the group's responses, which allows for a quantita- tive analysis and interpretation of data
- feedback of the results to the participating experts to inform them about the other experts' opinions, providing them with the opportunity for reflection – and, if applicable, reconsideration, clarification or change – of their views in light of the answers from the aggregate expert group
- iteration, which allows the experts to modify their views in light of the ag- gregate group's response through the feedback (see above) from round to round, which also allows for a gradual condensation of the general question

- a central monitoring group, which administers through a radial communication structure the data collection, the analyses, and the reciprocal information flow of the feedback

Through these features, it becomes apparent that the Delphi method includes elements of both quantitative and qualitative research, although it is listed among quantitative methods of data collection (Bortz & Döring, 2006, p. 261).

Häder (2009, p. 41) points out that cognitive aspects play a central role with respect to the underlying rationale of the Delphi method. Normative social influence and processes of group dynamics are regarded as a major influence on group judgment and decision making (Deutsch & Gerard, 1955; Woudenberg, 1991). In particular, in situations with unclear or complex issues, people may look to others as a source of information, which Deutsch and Gerard (1955) refer to as informational social influence. In this way,

> "[t]he Delphi technique was largely developed to avoid the problems of freely inter-acting groups, such as dominant individuals and pressure to conform to the majority view" (Bolger & Wright, 2011, p. 1500).

As outlined more precisely in 2.3.3.1.6, these sources of "process loss" are avoided by anonymization and controlled feedback (Bolger & Wright, 2011, p. 1500).

With respect to the cognitive background of participants' judgment formation in the first round of a Delphi study, Häder and Häder (2000, p. 24) refer to the information processing paradigm of personality (cf. Asendorpf, 2007), through which every individual can be described as an information processing system and which states that incomplete information can generally be supplemented by the knowledge of other individuals. Such processes occurs through active internal construction, for example in the form of a feedback process between perception, memory and information search. This notion leads to the assumption that iteration improves the quality of a Delphi study.

Bardecki indicates that "[t]he processes involved in an individual's response to the Delphi method are conditioned by a number of psychological effects" (1984, p. 281). As a major aspect, Bardecki identifies factors related to continuing participation in subsequent Delphi rounds (see 2.3.3.1.8).

In the following sub-chapters, central aspects with regard to the design of Delphi studies are outlined in more detail.

2.3.3.1.1 Initial Question and Question Types

A crucial aspect at the outset of a Delphi study is the operationalization of the issue. The facet theory of the social sciences provides a systematic approach to the preparation of the initial question of the study (cf. Borg, 1992). This theory is used to split a more general question or abstract issue into relevant dimensions (Häder, 2009, p. 88).

As it is generally assumed that with broad, open-ended questions, a wider range of responses can be gained than from a narrow set of questions, the context of the issue in a given Delphi study is usually structured into a qualitative first round and several, subsequent quantitative rounds (Häder, 2009, p. 87). The questions in the qualitative first round should be formulated on a more general level in order to provide the participants with a wide frame for their considerations. In this way, it can be avoided that the participants feel too restricted by guidelines (Häder, 2009, p. 116).

Throughout the study, the questions of the first qualitative round of a Delphi study develop into more focused and specific questions. With an iterative process like this (see 3.3.1.5), the experts' central opinions can be obtained and ideas in the course of subsequent rounds will be increasingly concretized and summarized (Häder, 2009, p. 116).

The decision for a particular question type is closely connected to the concept and purpose of the study. In general, there is a wide range of different question types that can be applied in a Delphi study. Question types which can be distinguished in the context of Delphi studies relate, for example, to choosing between different given possibilities, setting priorities, estimating consequences, pointing out alternatives, defining desirable und undesirable developments, exploring impacts, polling new needs etc. (Häder, 2009, p. 125). Engaging and concise formulations can entice potential experts' willingness to participate (Burkard & Schecker, 2014, p. 160; Skulmoski, Hartman, & Krahn, 2007, p. 10).

2.3.3.1.2 Expertise Criteria and Structure of the Sample

The deliberate selection and composition of the group of participants is of central importance with respect to the quality of the study, as the outcomes of a Delphi study are based on the participants' responses (Ammon, 2009, p. 464; Häder, 2009, p. 92). In this regard, Bolger and Wright (2011, p. 1507) point out that the process of a Delphi study is only successful if it involves a sufficient level of expertise in the panel. Hence, Delphi participants should meet certain expertise requirements. With the composition of the group of experts depending closely on the purpose of the study (Ammon, 2009, p. 464), Häder and Häder (2000, p. 18) point out that it is important to include experts with broad knowledge and comprehensive experience in the topic of the study so that a wide range of views and perspectives is represented. A determining factor for choosing experts is thus their affiliation with a context relevant for the issue of the given study (Ammon, 2009, p. 464).

Another important aspect is to ensure heterogeneity in the expert panel (Bolger & Wright, 2011, p. 1510; Linstone & Turoff, 1975a, p. 10). Hence, Delphi panelists should be chosen to represent different viewpoints (Ammon, 2009, p. 464), as in this way, "the likelihood that multiple frames on a situation will be generated within individual panelists is increased" (Bolger & Wright, 2011, p. 15101). Furthermore, the chosen Delphi participants should have the willingness, the capacity, the personal interest, and the endurance to participate throughout the different rounds of the study. An additional aspect is to focus on experts who are potentially able to disseminate and apply the results into practice within their context of influence (Häder, 2000, p. 18). Including experts with known "maverick" opinions could, moreover, enhance the process as they prompt and promote challenges to conventional thinking (Bolger & Wright, 2011, p. 1510). In general, it can be said that the selection of participants in a Delphi study is the determining and thus critical factor for the quality of its outcomes (Ammon, 2009, p. 466).

2.3.3.1.3 Sample Size

The question of sample size is closely connected to the context and specific purpose of a given Delphi study (Ammon, 2009, p. 465). Therefore, experiences with a large scale of different sample sizes in Delphi studies are reported (Häder & Häder, 2000, p. 18). As there are no general rules for an optimal size of a Delphi sample, the literature provides a variety of recommendations for the number of participants in a Delphi study (Häder, 2009, p. 96). Many recommendations range between 10 and 50 panelists (Häder, 2009, p. 96; Nworie, 2011, p. 26; Okoli & Pawlowski, 2004, p. 19). However, others point out that under appropriate conditions, even smaller groups of panelists can successfully participate in a Delphi study. Parenté and Anderson-Parenté (1987, p. 149) mention a number of ten participants as the minimum size of a Delphi panel. As a minimum size for sub-groups in differentiated analyses, researchers have suggested a critical threshold of seven participants (Becker, 1974, p. 12; Dalkey, Brown, & Cochran, 1969, p. 6). This number can also be linked to a theory known as *Miller's Law* from the field of information processing psychology, which suggests that the number of seven is a crucial element in a variety of applications in the context of dealing with complex systems and which implies that discussions with seven participants are most efficient (Miller, 1956). Nworie points out that smaller sample sizes are often recommended from a logistical perspective based on practical matters related to the coordination of Delphi study activities (2011, p. 26).

Many researchers assume that larger expert panels might help to reduce inaccuracy (Häder & Häder, 2000, p. 18). Accordingly, Ammon (2009, p. 465) points out that the more complex the topic of a given Delphi study is, the larger the size of the sample should be. Therefore, for Delphi studies with the aim of quantifying and qualifying opinions of experts, scholars recommend to involve as many participants as possible (Häder, 2009, p. 110). However, in the case of an all-qualitative Delphi study with the aim of determining opinions and arguments, fewer participants might be sufficient (Häder, 2009, p. 101).

A determining factor for the initial size of a Delphi sample is the estimated panel mortality (see 2.3.3.1.8). As the response rate usually diminishes throughout the different rounds, the initial sample should be large enough to provide a sufficient number of participants in the respective sub-groups in the last round after drop-out, thus promising sufficient data quality for interpretation (Ammon, 2009, p. 465).

2.3.3.1.4 Design of Questionnaires

With respect to designing questionnaires in Delphi studies, the specific characteristics of the Delphi method need to be considered (Häder, 2009, p. 122). In each round, a Delphi questionnaire should include information about the results of the corresponding previous round. As the survey takes place throughout several rounds, the questionnaires are labeled with identification numbers to allow for panel data analysis and response rate control. The participants of Delphi studies are not to be understood as test persons, but take on the role of informants. Thus, for the design of the questionnaires, it should be kept in mind that it is not the task of the participants to produce situation-based reactions. Rather, they are supposed to provide well-considered responses regardless of the survey situation. As opinion polling in a Delphi study generally takes place in written format, there is no need for taking into consideration elements of interviewer influence. In the quantitative rounds of Delphi studies, the questionnaires contain mostly standardized questions.

Usually, Delphi questionnaires are sent via mail or electronically. For paper-and-pencil based Delphi studies, a cover letter, a data protection statement, the questionnaire, a prepaid return envelope, and, if needed, additional information should be included. To obtain a high response rate, a number of recommendations with respect to design issues of questionnaires in Delphi studies can be retrieved from literature (Häder, 2009, p. 121). For the cover letter, an official letter head should be used. The purpose of the cover letter is to inform the experts thoroughly about the content and aims of the study. In particular, the purpose of the study and the importance of the experts' participation should be highlighted. Furthermore, the experts should be addressed personally and confidentiality should be assured. An appreciation of their efforts should be expressed as well. Also, it has been found that the length of the questionnaire has a negative effect on the response rate when exceeding approximately 12 pages. Furthermore, the questions and the structure of the questionnaire should be plausible. To facilitate the processing of the questionnaire, the questions should be arranged in a clear way and the sequence of the questions should be arranged in a top-down order. Questions related to the same issue should be grouped together into blocks of questions (Häder, 2009, pp. 122–124).

2.3.3.1.5   Iteration and Number of Rounds

As iterating the response process is a way of improving accuracy within a given group discussion (Häder, 2009, p. 207; Linstone & Turoff, 1975b, p. 234), Delphi studies are usually conducted as repeated interrogations throughout several rounds (Häder & Häder, 2000, p. 15). Bolger and Wright (2011, p. 1511) underline that it is especially "[t]he iterative feedback [that] sets Delphi apart from other nominal group techniques", as in this way, the Delphi method replaces direct confrontation and discussion by a systematic course of sequential individual interrogations through questionnaires, putting an emphasis on informed judgment (Bernice B. Brown, 1968, p. 7).

The number of rounds in a Delphi study is variable and depends on the purpose of the study as well as some pragmatic aspects. In general, it is argued that the validity of the results of the study is increased throughout the rounds (Linstone & Turoff, 1975b, p. 234). However, as the effort required of Delphi participants increases with the number of rounds, the response rate often diminishes over the rounds (see 3.3.1.8). Usually, the iterative process of polling and feedback is conducted until a predefined ending criterion is reached (Häder & Häder, 2000, p. 17). This can be, for example, the number of rounds, solidity of the results, consensus achievement, theoretical saturation, or when sufficient information has been exchanged (Häder & Häder, 2000, p. 119; Rowe & Wright, 1999, p. 355).

Reports of several applications of the Delphi method state that satisfying results were reached after three rounds (Häder & Häder, 2000, p. 17), as three rounds

"[m]ost commonly [...] proved sufficient to attain stability in the responses; further rounds tended to show very little change and excessive repetition was unacceptable to participants" (Linstone & Turoff, 1975b, p. 229).

For this reason, Delphi studies are typically conducted within three rounds (Häder, 2000, p. 17), with most of the changes taking place between the first and second round (Woudenberg, 1991, p. 140).

2.3.3.1.6 Anonymity of the Participants

The anonymity of the experts is an essential aspect with respect to the legitimization of the Delphi method. It is assumed that on the basis of anonymity, it is more convenient for the participating experts to revise their opinions without reservations, as an anonymous survey situation prevents them from a lost loss of prestige (Häder, 2009, p.148). Also, anonymity allows the experts to freely express their own opinions. It avoids influence through processes of group dynamics which can occur in common group processes as a source of inaccuracy and which can be induced by factors such as domination by quantity, group pressure towards conformity (Häder & Häder, 2000, p. 22), or the presence of opinion leaders (known as "bandwagon effect"), (Linstone & Turoff, 1975a, p. 4). Moreover, it can be assumed that an anonymous survey situation increases the experts' willingness to participate, as estimation and decision-making in complex issues can be related to some tentativeness or unease (cf. Kenis, 1995). Some researchers interpret anonymity among the participants in a Delphi study as a disadvantage, as participants in anonymous situations cannot be held responsible for their views and might in this way deliver blindfold or unelaborate statements. However, this assumption could not be empirically proven (Häder, 2009, p. 148).

2.3.3.1.7 Feedback and Structure of Information Flow

A further crucial aspect of the design of a Delphi study is the feedback of the results to the participating experts after each round. The feedback is directed by a central monitoring group and serves to inform the participants about the other experts' opinions. This provides them with the opportunity for a reflection – and, if applicable, reconsideration, clarification or change – of their views in light of the answers from the whole expert group. The feedback can be given in a variety of possible forms, including mean values, measures of dispersion, tables, graphs, statements etc. In case of verbal statements from the participants, common feedback strategies in the context of the Delphi method include an appropriate preparation and integration of such information into numeric feedback by the central monitoring group. In case of numeric assessments by the participants, statistical figures should be provided. The most common measure for the central tendency in the participants' assessments is represented by the arithmetic mean. Measures of dispersion are of particular importance in this case, as arithmetic mean values do not provide information about the variety and the distribution of the assessments. In such context, dispersion measures serve as indicators of the heterogeneity of the participants' views. A suitable measure for the distribution of responses emerges through the standard deviation (Häder, 2009, p. 151). In addition to the direct feedback, the researchers should provide the experts with access to detailed accounts of the results after each round and they should have the opportunity to ask the monitoring team for additional information (Häder, 2009, p. 156).

2.3.3.1.8 Panel Mortality

The retention of participants in a Delphi study is a crucial factor for the success of the study, as their commitment to the study throughout the different rounds is directly related to its success. Only if the knowledge of the participating experts can be continuously accessed from the first to the last round and is integrated into the feedback accordingly, optimal quality of the results can be expected (Häder, 2000 p. 19). Early withdrawal of experts can lead to information loss (Häder, 2009, p. 157).

The degree of experts' commitment to participating in a multi-round Delphi study is represented by the round-by-round response rate. The anticipated response rate is an important factor with respect to determining the size of the initial sample. However, the response rate in Delphi studies is difficult to estimate. In the first round of a Delphi study, usually about 30% of the contacted participants take part; in the following rounds, a response rate of about 70-75% can be expected. Previous experiences with the Delphi method show that high dropout rates occur especially between the first and second round of a Delphi study. As the elaboration on the tasks within a Delphi study usually requires much effort from the participants (Häder, 2009, p. 157), the response rate in Delphi studies also depends on the issue of the given Delphi study, the degree of its complexity, and the scope of the questions and tasks (Häder, 2009, p. 112).

Häder (2009, p. 158) lists three assumptions why experts abandon the Delphi process. The first refers to situations in which a participant's assessment deviates strongly from the other experts' opinions. In this case, elements of cognitive dissonance (cf. Festinger, 1962) might represent a motive for terminating participation (dissonance hypothesis). The second assumption relates to participants with more extreme assessments than the other experts. In this case, the large difference between the participant's own opinion and the other experts' views might cause the participant to leave the interrogation (non-conformity hypothesis). The third reason for participants' dropout from a Delphi process is connected to situations in which a participant provides his or her assessment with high uncertainty. The discontinuation of subsequent participation might in this case appear due a perception of their own incompetency (competence hypothesis). However, these hypotheses are not commonly agreed on and could not be confirmed empirically in several analyses. Therefore, dropout of Delphi participants could also be ascribed to other, more pragmatic factors such as a lack of time (Häder, 2009, p. 159).

A specific problem appears if, for example, systematic dropout of a certain sub-sample occurs. In this case, the views of a certain group would be underrepresented or lost. Hence, particular attention should be paid to non-response cases. This issue stresses the importance of sending additional reminders to the participants. Experiences indicate that after such reminders, approximately the same response rate occurs as in the corresponding original round (Häder, 2009, p. 159). For a reduction of dropout rates, sending reminders in time intervals of one, five, and seven weeks, and extending the deadlines, if possible, has proven successful (Häder, 2009, p. 123). In order to ensure that the targeted sample composition is reached, continuous response rate control by the monitoring team is recommended (Häder, 2009, p. 113).

2.3.3.1.9 Mode of Interaction

In Delphi studies, different modes of interaction are possible. In general, the communication throughout a Delphi study takes place in written form. This makes it possible to involve experts from different places. Initially, Delphi studies were conducted as paper-and-pencil versions with questionnaires being sent and returned by mail. In the 1970s, the application of real-time Delphi studies was tested in the USA, but due to the limited technological options during that time, it was only possible to involve a low number of experts. However, the advancement of communication technology provides new possibilities for the application of the Delphi method, for example, through electronic mail or web-based questionnaires. A great benefit of electronic communication in a Delphi study is that addressing the participants and the data collection can be directed in more efficient ways. However, the choice of interaction modes largely depends on the target group (Ammon, 2009, p. 464).

2.3.3.1.10 General Procedure

In the conventional Delphi process, the monitoring team designs a questionnaire which is sent to the anticipated respondent group (Linstone & Turoff, 1997a, p. 11). In general, the issue of the study is approached in a first qualitative round, which is followed by subsequent quantitative rounds (Häder, 2009, p. 87). The goal of the qualitative first round is the acquisition of a differentiated series of basic statements, representing a wide range of different views and thus avoiding one-sidedness of the study (Häder, 2009, pp. 116ff.). After the questionnaire is returned, the monitoring team analyses and summarizes the results, in the course of which the content of the responses is standardized in terms of language and is fed back to the participants. Based on these results, the monitoring team develops a questionnaire for the following round. In this way, the statements of the first qualitative round are subject to standardized assessment in the course of subsequent rounds. The participants are usually given at least one opportunity to reevaluate and refine their opinions in terms of ranking or rating their original answers based upon examination of the group response (Häder, 2009, p. 116; Linstone & Turoff, 1975a, p. 11).

2.3.3.2   Curricular Application of the Delphi Method

The Delphi technique can be effectively modified to meet the needs of a given study (Häder, 2009; Linstone & Turoff, 1975c). In investigations on educational issues, the Delphi method proves to be particularly suitable, as previously conducted curricular Delphi studies show (see 2.4). In curriculum related research, curricular Delphi studies in particular serve to gain insights in the context of determining educational goals, curriculum content and competence standards (Burkard & Schecker, 2014, p. 159; Häder, 2009, p. 238).

Following Frey (1980a, p. 32), the classical Delphi method as outlined above can be specified for curricular adaption by three additional elements. The first element relates to framing the overarching question with additional advice at the beginning of the interrogation and specifying aspects to induce reflection and at the same time to avoid stereotypical answers. The second element refers to collecting views and opinions within a curricular structure that is specified by different curricular dimensions such as contexts and situations, topics and concepts, and competences and attitudes (cf. 2.2.2). This element systematizes the interrogation, ensures that the statements of the participating experts relate to potential curricular situations and thus avoids a collection of only broad and common objectives, terminology, and domains. In this way, the general question is specified within a formal question and answer format. The third aspect refers to criteria for selecting experts involved with curricular matters, as being part of society does not inevitably qualify for being considered as an expert in the field of curriculum research. Thus, selecting participants on the basis of these criteria ensures a group of experts that is legitimated from a curricular perspective.

In the curricular modification of the Delphi method, the focus is on normative and subject-specific considerations of education (J. Mayer, 1992, p. 97).

With this purpose, a curricular Delphi study inquires of a wide range of opinions. As divergence can occur in interrogations that contain issues of value conflicts such as in the field of education, it is possible that instead of convergence, dissent appears within this range of views. However, as Mayer (1992, p. 97) points out, dissent regarding educational concepts does not rule out a consensual acceptance of the validity of such different opinions, as the discourse does not aim at a consensus of values but at consensual reflection on the legitimacy of different values.

2.3.4 Summary

A curriculum process is constituted by three elements. These refer to focus and objectives, interaction and discourse, and reflection. Following Frey (1980b, pp. 27–28), the curricular process carried out within this study for answering the question of desirable science education in terms of scientific literacy is based on the following assumptions:

- The question of desirable aspects of science education cannot be answered by one expert alone, but different experts are necessary who feature certain characteristics. Generally, the experts should comply with the overarching objectives of education in general and science education in particular. Moreover, the groups of experts should be competent with respect to the question of desirable science education.
- The curricular process of inquiring, collecting answers, providing feedback etc. is to be designed in such a way that it supports enlightening and systematic reflection.
- The experts engaged in this curricular process should interact with each other through systematic reflection.
- The outcomes of the curricular process should provide the possibility for further elaboration and adaption to other persons involved with curricular matters of science education as well.

Therefore, a crucial factor in a curriculum process is the involvement of appropriate stakeholders. As relevant stakeholders in the context of desirable aspects of science education, students with different science subjects, science teachers (including science education students at university, trainee science teachers, inservice science teachers, and science teacher supervisors), science education researchers, scientists, and representatives from education policy and administration can be identified. These stakeholder groups represent experts to be included in a reflection on the curricular content of desirable scientific literacy based science education.

An appropriate way of collecting and structuring experts' opinions on educational issues is represented by the Delphi method, which has already proven suitable to compile views on complex issues such as contexts, contents and aims of education. The Delphi method is a specific form of structured group communication and is used for accessing the expertise of numerous persons with different competent perspectives on a given issue. A particular advantage of the Delphi method for the purpose of this study is the anonymity of the participants. As outlined in chapter 2.3.3.1.6, ensuring respondents' anonymity avoids several negative effects such as influence by opinion leaders that can emerge from direct group interaction and face-to-face discussion. Another significant aspect is the iterative and controlled feedback, which promises greater accuracy and thus more solid results. Furthermore, the open initial situation question is an important factor, as it helps to avoid narrow initial responses. Through the written form of communication, a high number of experts from different places can be involved. This is of particular importance both with regard to the international framework which this study is part of and with respect to subsequent meta-analyses. The curricular adaption of the Delhi method includes three additional elements. First, the overarching question at the beginning of the study is supplemented with additional advice and specifying aspects to initiate reflection and avoid stereotypical answers. Secondly, views and opinions are collected within a curricular structure specified by different curricular dimensions such as contexts and situations, topics and concepts, and competences and attitudes. This is done in order to systematize the interrogation, obtain statements relating to potential curricular situations and avoiding broad and common objectives, terminology, and domains. The third element refers to criteria for selecting experts dealing with curricular issues and in this way being legitimated from a curricular perspective.

2.4 Science Education in Light of Subject Specific Curricular Delphi Studies

Various studies have been conducted with the aim of gathering and exploring views of various groups of stakeholders on specific issues of science education. The Delphi method has frequently been used to investigate and clarify curricular aspects regarding several sub-domains or subject-specific issues of science education (Bolte, 2003a, 2003b, 2008; Edgren, 2006; Häußler et al., 1980; Heimlich et al., 2011; J. Mayer, 1992; Osborne et al., 2003; Welzel et al., 1998). In the following, I will describe three examples of subject-specific curricular Delphi studies in the context of science education in Germany. The studies explore the views of several stakeholders from different domains associated with school science and focus on desirable aspects of biology, physics, and chemistry education respectively as part of general education.

2.4.1 Physics

The first curricular Delphi study in science dates back to the 1980s and was conducted by Häußler and his colleagues (1992; 1980). The study focused on "physics education of today and tomorrow". The authors of the study used the curricular Delphi method (see 2.3.3.2) to determine aspects of desirable physics education, drawing on the expertise of an initial sample of 73 stakeholders selected according to specified criteria. The stakeholders represent different groups associated with physics and physics education, e.g. physics teachers, physics education researchers, physicists, employees in physics-related industry or other related contexts, education policy, members of curriculum committees, and general educationalists (Frey, 1980c, pp. 38–41). Within three consecutive rounds, these stakeholders were asked for their opinions on the question "what should physics education look like so it is suitable for someone living in our society as it is today and as it will be tomorrow?" (Häussler & Hoffmann, 2000, p. 691). The first round gathered the participants' initial views on this issue in an open questionnaire format that was divided into the curricular domains of contexts and situations, topics and concepts, and qualifications. In the second round, the degree of priority and realization in the classroom of the collected aspects was determined. Also, the participants developed more particular statements regarding desirable physics on the basis of the given aspects through assembling combinations from the given categories. Moreover, the relevance of physics education was assessed in relation to other fields of education in a third task. Through a hierarchical cluster analysis, the category combinations of the participants were merged into concepts of desirable physics education. These concepts were subject of the third round and were again assessed by the participants. Here, the assessment focus was on the concepts' importance and the degree of their realization in practice as well as further aspects of the initial research question. Moreover, the participants were asked to outline content-based specifications of a given group of themes for educational situations on the basis of the concepts (Häußler et al., 1980).

The results of the qualitative round of this study yielded 54 aspects that were assigned to an either scientific, professional, cultural, social, or personal domain. Contextual categories of desirable physics education included, for example, the scientific structure of physics, insights into the professional world, basic qualification for careers, understanding of the implications of scientific and technical developments and the risks associated with it, avoiding safety hazards and accidents in daily life, household and living environment, leisure, society and public, consumer behavior, emotional personality area, gratification in dealing with science, intellectual personality area, and general education (Häußler & Rost, 1980a; Hoffmann & Rost, 1980)

With respect to desirable content of physics education, topics related to "scientific knowledge and methods as mental tools", "passing on scientific knowledge to the next generation", "physics as a vehicle to promote practical competence", and "physics as a socio-economic enterprise" were given highest priority (Häußler & Hoffmann, 2000, p. 693). This shows that the participants saw physics "more as a human enterprise and less as a body of knowledge and procedures" (Häußler & Hoffmann, 2000, p. 704). In general, the outcomes of the priority assessments suggest some degree of consensus among the participants concerning the different aspects of cultural and social relevance of physics education, the knowledge domains, and competences. As for the significance of physics education for the individual, about half of the participants emphasized the cognitive and emotional relevance. The other half stressed aspects relating to practical issues of everyday life. The practice assessments showed that in most of the cases, the practice assessments fell short of the priority assessments. Particularly high deficiencies, indicated through the priority-practice differences, were attributed to competency aspects such as problem solving, socio-political acting, self-reflection, and applying knowledge (Rost & Spada, 1980, pp. 188–191).

The five concepts obtained from cluster analysis of the category combinations relate to "responsible socio-political acting and public discussion of physics and technology" (1), "mastering and understanding physical and technical devices" (2), "enhancement of the emotional experience of nature and technology and satisfying occupation with physics" (3), "promotion of an occupation with the scientific tradition of physics as an intellectual endeavor" (4), and "insights into the professional domain and qualification for professional life" (5) (Häußler & Rost, 1980b, p. 220). According to the priority and practice assessments of the concepts, the largest perceived deficiencies appeared for the concept related to responsible socio-political acting and public discussion of physics and technology (1). This finding can be understood as a clear vote by all participating experts for a stronger orientation of physics education towards the requirements of responsible participation of the individual in society and public debate (Rost & Spada, 1980, p. 254).

As the sample of the curricular Delphi study in physics does not include students as stakeholders for physics education, views of about 500 students on the issues investigated in this study were addressed in a follow-up study, investigating their interests in the contents, contexts, and activities that were suggested by the curricular frame of this Delphi study (cf. Hoffmann & Lehrke, 1986). In this follow-up study, a "remarkable congruency between students' interest in physics and the kind of physics education identified in the Delphi study as being relevant" could be identified. In addition, „a considerable discrepancy between students' interest and the kind of physics instruction practiced in the physics classroom" was determined (Häußler & Hoffmann, 2000, p. 689). Moreover, the results of the follow-up study showed that similar to the priority-practice differences determined in the curricular Delphi study on physics, mismatches for the concepts from the curricular Delphi study on physics appeared between the interests of the students and the degree of their perceived realization in the classroom. In this way, the results suggest that an orientation of physics education towards the particular concepts may be crucial in terms of students' interests (Häußler, 1992, p. 139).

2.4.2 Biology

Following the work of Häußler et al. (1980), Mayer (1992) conducted a Delphi study on the issue of biodiversity in the context of compulsory biology education. Tying in with the notion that the task of didactic discourse is to make choices about educational contents that should be considered relevant with respect to general education, the study focused on the overarching curricular question of which contents of biodiversity are relevant and pedagogically desirable for the individual in the society of today and the near future. The responses to this question were compiled in a curricular frame that involved educational goals, methods, and relevant content related to biodiversity such as biotopes, biocenoses, and living beings (Mayer, 1992, p. 12).

In this study, the views of an initial sample of 77 participants were collected in three rounds. The sample consisted of different stakeholders that qualified as experts with respect to the question of the study. These included biology teachers, biology education researchers, educators, biologists in research, industry, and administration, students at the level of upper secondary education, and representatives from out-of-school educational contexts. Starting with an open three-part questionnaire, the overarching question is in the first round were specified into contexts (I), content (II) and competences (III), as done in the Delphi study in physics by Häußler and his colleagues (1980). In the second round, a synopsis of the first round took place through the participants' combinations of categories derived from the responses in the first round. These were merged into meaningful statement complexes, and category frequencies were determined. The category combinations were further specified by the participants with respect to particular organisms and biotopes. Concepts of teaching topics related to biodiversity were derived through hierarchical cluster analysis of the category combinations as well as from the content specification provided by the participants, In the third round, these concepts were assessed by the participants according to their priority, and further opportunities for extending the lists of relevant species as well as selecting the most important ones are provided (p. 99).

With respect to contexts and motives of teaching content related to biodiversity, 15 different categories were determined. The most frequently mentioned aspects with respect to the issue of the study included the protection of endangered species and dealing with environmental problems, an enlightened appreciation of nature, an understanding of the scientific tradition of biology, an emotional relation to nature, and coping with life. The question about desirable content related to biodiversity yielded 17 thematic areas, referring to systematic, morphological, ecological, recreational, and application based aspects. Special emphasis was placed on ecological considerations and students' everyday life. With regard to the question of competences to be enhanced through biodiversity related content, 15 aspects of knowledge, skills, methods, and attitudes were determined. In their assessment, the participants placed great emphasis on the knowledge of names and manifestations of animals and plants, the classification and systematization of species, a critical attitude towards human treatment of nature, and protective and responsible behavior regarding living beings. The cluster analysis yielded five concepts with regard to addressing biodiversity in the classroom. These include "living beings within the context of ecology and environment protection" (1), "living beings as part of general biological and physiological manifestations" (2), "dealing with living beings from the perspective of diversity of organisms and their systematization" (3), "occupation with living beings in the context of leisure and experiencing nature" (4), and "living beings in light of advantages and harms for humankind" (5). In general, the results show that teaching about biodiversity can take place within numerous topics of biology education. Moreover, Mayer argues that addressing biodiversity should also be considered as an overall task of biology education. In conclusion, he points out that dealing with biodiversity in the biology classroom should not take place in an isolated way from other content, but in combination with general and applied biology and along descriptive and analytical methods (pp. 272–275).

2.4.3 Chemistry

Following the approaches by Häußler and his colleagues (1980) and Mayer (1992), Bolte (2000, 2001, 2002, 2003a, 2003b, 2008) used the Delphi method to determine aspects of meaningful and pedagogically desirable chemistry education considered relevant for the scientifically literate individual in the society of the present and the future. For this purpose, the study analyzed dimensions as well as fields of dissent and consensus in the opinion of 114 stakeholders. The particular aim of the Curricular Delphi Study in Chemistry was to facilitate

> "a reflection on content, task, and aims, as well as the development of guidelines for a modern scientific literacy based – chemistry-related – basic science education from different stakeholders' views" (Bolte, 2008, p. 334).

Collecting the views of students, chemistry education students at university, trainee and in-service chemistry teachers, trainee teacher supervisors, chemistry education and science education researchers, representatives of chemistry teacher associations, chemists, and other professionals from science-related fields, the study covered different groups that are affected by chemistry-related science education (Bolte, 2003b, p. 12).

In the first round, these experts provided their opinions on aspects of desirable chemistry education in an open questionnaire, which was structured according to contexts, situations, and motives (I), content (II), and qualifications (III). Their statements were analyzed by means of qualitative and quantitative methods and were summarized into categories. These categories were reported back to the participants for weighted assessment in the following round. In light of the general opinion of the expert panel, the participants assessed the priority and the extent to which these aspects were realized in practice. This process revealed which characteristics featured higher and which featured lower importance in the participants' opinions. On the basis of these assessments, priority-practice differences were calculated to identify especially deficient areas in chemistry education and the degree of consensus among the stakeholder groups about assessing these aspects was determined. In addition, with the aim to identify concepts of desirable chemistry education, the participants were asked to compile meaningful category combinations. Identified through cluster analytical processes, these concepts were fed back to the participants for weighted evaluation analogously to the second round and for further assessment in the third round (Bolte, 2003b, pp. 9–10, 2008, pp. 335–336).

In the results of the first round, general trends as well as specific emphases can be identified in the opinion of the participants. The analysis of the participants' statements yields a total of 60 categories of desirable chemistry education. As expected by the author of the study, the identified aspects coincided to a great extent with those criteria included in recommendations for modern chemistry education by didactic and teacher associations, but also enhanced those recommendations with further aspects (Bolte, 2003b, p. 13). Moreover, the results of the first round revealed a shift of focus in the context of desirable and modern chemistry education, pointing to the importance of embedding chemistry based topics in everyday life related contexts, interdisciplinary or multi-perspective approaches to topics, and the consideration of qualifications that provide the basis for lifelong learning in chemistry. This shifted focus was also supported by the quantitative analyses (Bolte, 2003b, p. 14). They revealed that all participants placed high emphasis on aspects related to household, everyday life, the environment, basic knowledge, nutrition, health, sustainability, relation of chemistry to social issues, historical perspectives, scientific chemistry-related developments, understanding, judgment abilities, and action competence (Bolte, 2003b, pp. 15–17).

With respect to the category frequencies of the four sub-samples of this study (students, teachers, education researchers and scientists), the differentiated analyses support the author's initial consensus-dissent-hypothesis, which assumed a gap between the expectations of science education and the educational interests of large sections of the population (Bolte, 2008 p. 333). More specifically, the results show that on the one hand, there seems to be a general consensus among the participants on several content related aspects of science education, but on the other hand, the four different sub-samples show different accentuations in their views on meaningful and pedagogically desirable aspects of chemistry related scientific literacy of the individual in the society of today and tomorrow (Bolte, 2003b, p. 20). Hence, the results suggest that

> "the much praised consensus about the importance of scientific education and about how this education should be realized, is seemingly inappropriate" (Bolte, 2008, p. 341).

In particular, Bolte found that the group of young people and the three adult sub-samples expressed different ideas. This finding supports what Bolte calls the "hypothesis of the educational conflict of the generations". This hypothesis relates to the notion that science curricula are mostly developed by experts who are scientifically socialized, as well as the fact that the content of science lessons is for the most part determined by the teachers – whereas students do not necessarily share the opinions and priorities of these groups. With respect to educational intentions and educational offers, chemistry education is thus "dominated by adults' conceptions of good general education, whereas young people's educational interests remain ignored" (Bolte, 2008, p. 333). In the Curricular Delphi Study in Chemistry, this discrepancy is particularly shown by contrasting the students' views with those of the scientists. As these differences are primarily the result of certain aspects being hardly mentioned by the group of students in the first round, Bolte moreover raises the question whether this might be related to the fact that the these characteristics appear only barely or implicitly in common chemistry education, which leads to the "versatility-versus-one-sidedness-hypothesis" (Bolte, 2008, p. 341). He argues that

"if science lessons are primarily planned and held according to the structure that the pure subject lays out (science first), and problems in society or in the world are only dealt with afterwards [...], then it is obvious that there is an imbalance between the central intentions of chemistry related to formal education (personal relevance first) and chemistry related specialization undertaken in schools" (Bolte, 2008, p. 333).

On the basis of the iterative process of the Delphi method, more solid claims about this assumption were retrieved from the results of the second round. The results of the second round of the Curricular Delphi Study in Chemistry show that out of the given 60 categories from the first round, only three aspects were assessed as not so important (priority mean ≤ 2.75), all other 57 categories were assessed as important or very important. This indicates the validity of the classification system. From the practice perspective, however, only eight of these aspects were assessed as being present in chemistry lessons. These categories refer to aspects that relate to chemistry as a science and structure-of-the-discipline approaches (Bolte, 2008, p. 341). Categories representing everyday life or nature of science-oriented aspects were, on the other hand, assessed as sparsely present in chemistry lessons. Furthermore, as almost all categories featured large priority-practice differences, the results show that according to the participating experts, chemistry education needs improvement (Bolte, 2008, p. 342). Areas identified by the experts as most deficient and thus featuring the most urgent need for change include aspects referring to motivation and interest, value systems, judgment ability, reflected action, and multi-disciplinary approaches (Bolte, 2008, p. 344). The results of the second round also show that the three hypotheses can statistically not be falsified. Moreover, the quantification of the second round revealed several aspects that could generate and enhance discussions about chemistry as an essential part of general education (Bolte, 2008, p. 343).

All in all, the outcomes indicate that chemistry related science education needs reform. In particular, chemistry education in practice is still found to be characterized by an overemphasis on aspects related to chemistry as a science, and less oriented towards issues of general education, as demanded in frequent discourse on scientific literacy. As a result, chemistry teachers have to take into account that the perception of current conventional chemistry lessons does – as assessed by the experts in the Curricular Delphi Study in Chemistry – not coincide with desirable chemistry education. This applies both to the adult stakeholder groups and the group of students. Furthermore, Bolte has pointed out that "to enhance scientific literacy, it is necessary to have both sides in mind and to focus on both, the educational expectations of society (or of the 'adults') and the educational interests of the younger generation", negotiating aims and topics of chemistry related science education with both students and adults as representatives of society. The findings of the Curricular Delphi Study in Chemistry can help to bridge the gap between these groups (Bolte, 2008, p. 344). Considering scientific literacy as a major aim of science based general education, the results furthermore imply that in chemistry related curricular discourse, it is crucial to clarify the genuine contribution of chemistry to a modern worldview and to explain which scientific-cultural advancements the prosperity and quality of life in industrialized societies is based on. This should be embedded in an understanding that it can be personally fulfilling to be able to reflect on problems of one's own living environment from the perspective of chemistry (Bolte, 2003b, p. 24).

2.4.4 Summary

The notion of the need to include members of society in a discourse on curricular aspects of science education has generated several studies which address the views of different stakeholders. The curricular Delphi method has been established and applied as a particularly appropriate method for addressing curriculum related issues within a structured group discussion. The three curricular Delphi studies outlined above follow the same general design: A fixed group of participants is interrogated about aspects of desirable aspects of physics, biology, and chemistry related science education throughout three consecutive rounds. They are selected on the basis of specified criteria according to their science education related backgrounds and include students, teachers, education researchers, scientists, and representatives from education associations and education policy. For ensuring anonymity among the participating stakeholders, their views are collected and analyzed by a central working group, which administers the information flow. In a first round, their views are collected in an open questionnaire and classified into categories. In a second round, the results from the first round are fed back to the participants for further (quantitative assessment) allowing for reconsideration of their own opinions in light of the group opinion. In a third round, concepts of science education with respect to the subject-specific focus of the corresponding study derived through cluster analyses of category combinations from the second round are fed back to the participants again for weighted assessment. All three studies point out general tendencies as well as specific insights with regard to their particular foci of science education and reveal a need for action in science education within their field of investigation, providing specific starting points and orientation frameworks for improving physics, biology, and chemistry related general education.

2.5 Research Questions and Hypotheses

In the preceding chapters, essential aspects that are connected to processes of determining scientific literacy based science education were addressed. On the basis of the theoretical reflections, the need for a comprehensive approach towards desirable scientific literacy based science education from the perspective of different stakeholders in society was pointed out. For this reason, the main aim of the Curricular Delphi Study in Science is to investigate the views and opinions about desirable aspects of scientific literacy based general education of affected members of society such as students, teachers, science education researchers, and scientists. The central question of this study is:

What aspects of science education are by different stakeholders considered meaningful and pedagogically desirable for the scientifically literate individual in the society of today and tomorrow?

This question is specified within several research questions and investigated on the basis of the corresponding hypotheses. To find answers to this question, a suitable method for collecting and processing different stakeholders' views is needed. As shown by several scholars, the Delphi method has proven to be a suitable instrument to investigate stakeholders' opinions on curricular aspects in different science subjects (Bolte, 2003a, 2003b, 2008; Häußler et al., 1980; J. Mayer, 1992; Osborne et al., 2003).

Based on the works of Bolte (2003a, 2003b, 2008), Häußler (1980), and Mayer (1992), I expect that through the Delphi method, a valid classification system for desirable aspects of science education can be obtained. As in previous Delphi studies of science education and with reference to widely acknowledged aims and objectives of science education, I expect that several aspects expressed by the stakeholders of this study relate to recommendations in literature. Furthermore, based on sample and methodological similarities, I expect that the views provided by the stakeholders of this study relate to aspects collected in previous Delphi studies. However, in view of the integrative approach to science education in this study and the large time interval between this study and the aforementioned previous Delphi studies, I also expect that the statements provided by the stakeholders of this study include additional aspects. On the basis of these considerations, I address the following research question and hypotheses.

1. What expectations of desirable science education can be identified in the stakeholders' views?

Hypothesis 1a: Through the curricular Delphi method, a valid classification system for desirable aspects of science education can be reached.
Hypothesis 1b: The aspects of desirable science education expressed by the participants of this study relate to recommendations from literature
and to aspects collected in previous curricular studies associated
with science education, but also include additional aspects.

Following Bolte (2003a), I expect that the stakeholders assign highest priority to
aspects related to scientific inquiry, environmental issues, content knowledge,
and overarching aims of general education. As well, I expect on the basis of the
results obtained by Bolte that, in contrast, the stakeholders give lowest priority to
aspects connected to the structure of the science disciplines, specialized fields,
and traditional approaches of single subject orientation. Taking into account the
progression of complexity in science education towards higher levels of education (KMK, 2005a, 2005b, 2005c), I assume that the priority of all aspects increases with more advanced levels of education.

    Based on the results by Bolte (2003a) with respect to the perceived presence
of aspects desirable science education in the science classroom, I expect that
aspects related to the structure of the science disciplines, specialized fields, and
traditional approaches of single subject orientation are assessed with the highest
extent of realization. Following Bolte (2003a) and based on frequently demon-
strated failures of science education to meet overarching aims of education (e.g.
Deutsches PISA-Konsortium, 2001; OECD, 2004a, 2007b, 2010), I anticipate
that the stakeholders assign lowest degrees of realization to aspects related to
interdisciplinarity, students' living environment, ethical references, and overarching aims of general education. Taking into account recommended progres-
sions in science education such as expressed in the national education standards
(KMK, 2005a, 2005b, 2005c), I expect that the extent of realization of all aspects
increases with more advanced levels of education.

    According to Bolte (2003a), a gap between science educational practice and
the educational interests of large sections of the population can be identified.
More specifically, for most aspects considerable discrepancies between stakeholders' priorities of desirable chemistry related science education and their
actual perception of reality of appear in terms of an underrepresentation of these
aspects in practice. Therefore, I expect that for most of the aspects of desirable
education that are identified throughout this study, substantial priority-practice
differences appear in terms of their perceived realization not living up the stakeholders' priorities. Following further results obtained by Bolte (2003a), I assume
that most of the largest priority-practice differences appear for aspects considered as most important by the stakeholders. These expectations are specified on

the basis of the following research questions and hypotheses, which allow for statistical evidence whether these expectations can be confirmed.

2. What emphases can be identified in the stakeholders' views?

Hypothesis 2a: Highest priority is given to aspects related to scientific inquiry, environmental issues, content knowledge, and overarching aims of general education.

Hypothesis 2b: Lowest priority is assigned to aspects connected to the structure of the science disciplines, specialized fields, and traditional approaches of single subject orientation.

Hypothesis 2c: The priority of aspects depends on the level of education and increases with more advanced levels of education.

3. To what extent are aspects of desirable science education realized in practice according to the stakeholders' views?

Hypothesis 3a: Aspects related to the structure of the science disciplines, specialized fields, and traditional approaches of single subject orientation are assessed with the highest extent of realization.

Hypothesis 3b: Aspects related to interdisciplinarity, students' living environment, ethical references, and overarching aims of general education are assessed with lowest extent of realization in current science education.

Hypothesis 3c: The extent of realization depends on the level of education and increases with more advanced levels of education.

4. What differences between priority and extent of realization regarding aspects of science education can be identified in the opinions of the stakeholders?

Hypothesis 4a: For the majority of the aspects, considerable priority-practice differences appear.

Hypothesis 4b: For most aspects, their realization falls short of their priority.

Hypothesis 4c: For the highly prioritized aspects, especially large priority- practice differences can be identified.

Differentiated analyses of the curricular Delphi study in chemistry (Bolte, 2008 p. 333) provide insights about whether the sub-samples share some consensus or drift apart in their opinions (consensus-dissent-hypothesis). On the one hand, there seems to be some general consensus among the participants on aspects of desirable chemistry related science education. However, on the other hand, the four sub-samples feature different emphases in their views (Bolte, 2003b, p. 20). This points to several distinctions in the stakeholders' opinions about meaningful and pedagogically desirable aspects of chemistry based scientific literacy of the individual in the society of today and tomorrow and suggests that the frequently claimed consensus about the how to realize scientific literacy based science education best with respect to chemistry education is not appropriate (Bolte, 2008, p. 341). Following these findings, I expect that those aspects considered most desirable for science education by the sub-samples in this study broadly correspond to each other, but however, the sub-samples differ in their specific assessments of aspects of desirable science education.

The Curricular Delphi study in Chemistry (Bolte, 2008, p. 333) furthermore reveals that particularly the group of young people and the three adult sub-samples express different ideas (hypothesis of the educational conflict of the generations). Bolte relates this finding to the notion that science curricula are mostly developed by experts who are scientifically socialized and that science lessons are mostly determined by teachers, whereas students do not necessarily share the opinions and priorities of these groups, so that chemistry education is mostly dominated by adults' conceptions of good science education, whereas young people's educational interests remain ignored. As today, science education is still seen as usually defined by the academic community (Nixon et al., 1996, pp. 270, quoted in Jenkins, 2006, p. 1; Osborne & Collins, 2001, p. 442; Osborne et al., 2003, p. 693), I expect that in the Curricular Delphi Study in Science, a gap between the assessments of the young generation and the adult groups regarding aspects of desirable science education appears.

These expectations are specified by the following research question and hypotheses.

5. Is there a consensus among the different stakeholder groups regarding their opinions assessments of aspects of desirable science education?

Hypothesis 5a: The aspects considered most desirable for science education by stakeholders of the different groups broadly correspond to each other.

Hypothesis 5b: However, the stakeholder groups differ in their specific assessments of aspects of desirable science education.

Hypothesis 5c: There is a particular gap between the assessments of the young generation and the adult groups regarding aspects of desirable science education.

3 Method

The considerations regarding the investigation of desirable aspects of scientific literacy based science education outlined in the theoretical part are taken account of by the methodical outline of this study, which is addressed in the following sections. First, I will describe the design of this study. In the second chapter of this part, I will, according to the different rounds of this study, outline the design of the questionnaires and describe the methods applied to analyze the data.

3.1 Design of the Berlin Curricular Delphi Study in Science

For engaging in a process with the aim of developing a curricular framework, the selection of the general methodology of the study and the composition of the sample are of particular importance (cf. 2.3.1 and 2.3.2). These aspects are addressed in the following sections of this chapter.

The considerations in this part are made with respect to the Berlin Curricular Delphi Study in Science. However, as this study is part of the International PROFILES Curricular Delphi Study on Science Education, it is one of several independently realized national Delphi studies carried out within the frame of the PROFILES project in different countries (cf. 2.1.5).

With Freie Universität Berlin as the leader of the work package in which these Delphi studies are embedded (work package 3: stakeholder involvement and interaction), general design considerations of the Berlin Curricular Delphi Study in Science as outlined in the following sections are also followed by the majority of the other PROFILES partners' Delphi studies (Bolte, Holbrook, Mamlok-Naaman, & Rauch, 2014; Bolte & Schulte, 2014a, 2014b, 2014c; Bolte, Schulte, Kapanadze, & Slovinsky, 2012; Börlin & Labudde, 2014; Charro, Plaza, & Gómez-Niño, 2014; Gauckler et al., 2014; Kapanadze & Slovinsky, 2014; Keinonen, Kukkonen, Schulte, & Bolte, 2014; Ozdem & Cavas, 2014; Rundgren, Persson, & Chang-Rundegren, 2014; Schulte, Bolte, et al., 2014; Schulte & Bolte, 2012, 2014a, 2014b; Schulte, Georgiu, et al., 2014).

3.1.1 Structure and Procedure of the Study

As discussed in 3.3, the Delphi technique appears to be a suitable instrument to investigate curricular aspects of science education. Following Frey (1980a, p.

31), the question of desirable aspects of science education is addressed best by applying the following elements of the Delphi method:

-   involving a number of selected stakeholders (experts) in the field of science education
-   iterative interrogation through formalized questionnaires
-   consecutive feedback of the answers of the whole group
-   a formalized question format that starts with an open question, which is then condensed and concretized throughout the following rounds
-   administration of the processes by a central working group

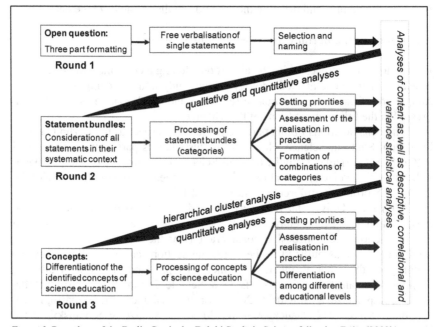

Figure 1. Procedure of the Berlin Curricular Delphi Study in Science following Bolte (2003b)

From these elements and the theoretical considerations in 2.3.3.1, several implications emerge for the design and methodical concerns of the Berlin Curricular Delphi Study in Science. The procedure of this study is shown in Figure 1.

Following recommendations from literature (Linstone & Turoff, 1975b, p. 229; Häder, 2000, p. 17; Woudenberg, 1991, p. 140) and the design of previous curricular Delphi studies on different science subjects (cf. 4.2), this study is conducted in three rounds. In addition, the Berlin Curricular Delphi Study in Science follows the typical procedure of structuring a Delphi study into a qualitative first round and subsequent quantitative rounds (Häder, 2009, p. 87), which has proven suitable in curricular application of the technique (cf. 2.4). The communication and information flow is managed by a central working group.

The *first round* aims at accessing the range of opinions of the stakeholders involved in this study. For this purpose and following Bolte (2003b), the participants' views about desirable aspects of scientific literacy oriented science education are first collected through a three-part open questionnaire (see 3.2.1). All stakeholders in this study are presented with the same questionnaire. In the course of the qualitative and quantitative analyses of the first round, the participants' individually formulated answers are organized, labeled, summarized into categories (statement bundles) and statistically evaluated by determining category frequencies and testing for differences (3.2.1.3). In this way, the statements of the first round are consolidated into meaningful categories.

As the questionnaire for the *second round* is in Delphi studies usually generated from the results of the first round, these categories are in the second round of this study reported back to the same participants for weighted assessment.

A crucial factor to improve the quality of the results is continuous verification throughout the Delphi process. In this way, the stakeholders have the opportunity to verify in light of the general opinion of the expert panel whether their responses in the first round indeed reflect their opinions or to reconsider, clarify, change, or expand their answers from the first round, and a Delphi-specific gradual condensation of the general question is pursued (Häder, 2009, p. 25; Häder & Häder, 2000, p. 17).

More specifically, following Bolte (2003a), the participants are asked to prioritize the given categories and to assess to which degree these categories are realized in practice. This procedure reveals which aspects the participants consider of higher or lower importance. Also, this can show whether aspects that were very frequently or rarely mentioned in the first round are considered especially important or unimportant, or if the distribution of the frequency patterns is based on other than priority reasons. Moreover, on the basis of these assessments, priority-practice differences (cf. Bolte, 2003a) are calculated to identify areas in science education that participants consider especially deficient (see 3.2.2).

In addition, the participants are asked to compile meaningful combinations from the given set of categories. Through hierarchical cluster analysis, these combinations are merged into concepts of desirable science education (cf. Bolte, 2003a). In this way, the categories from the second round are further processed and condensed towards more abstract and broader conceptual statements.

In a final *third round*, the identified concepts of desirable science education are reported back to the participants from the second round for assessment according to priority and degree of realization in practice, analogous to the second round. In a further step, the participants are also asked to differentiate their assessment for different educational levels. In this manner, the third round continues the process of continuous specification of the overarching question on the basis of the responses from the previous round.

In general, the questions of the first qualitative round of this study lead to more specific questions throughout the Delphi process. With such an explorative procedure, experts' central opinions on desirable science education can be extracted and in the subsequent rounds more and more concretized and summarized (Häder, 2009, p. 116).

In order to include both conventional written forms of communication and new possibilities which emerged through the advancement of internet technology, this study applies a two-pronged approach of interaction and communication. One the one hand, a paper-and-pencil questionnaire is used for situations such as students and teachers filling in the questionnaires at school. On the other hand, a web-based questionnaire is developed and personalized access codes are distributed via email to enable more flexible participation regardless of place and time.

3.1.2 Selection of the Sample

As a Delphi study is only successful if it involves a sufficient level of expertise in the panel, the composition of the group of participants is of central significance. Therefore, Delphi participants should fulfill certain expertise demands with respect to the topic of the given study and feature an affiliation with a context relevant for the issue of the given study (Ammon, 2009, p. 464; Bolger & Wright, 2011, p. 1507; Häder, 2009, p. 92).

With the purpose of this study to capture a wide range of views and perspectives as well as to establish a synopsis of general opinions, an important aspect is to ensure heterogeneity in the expert panel and thus to choose participants who represent different viewpoints (Ammon, 2009, p. 464; Bolger & Wright, 2011, p. 1510; Linstone & Turoff, 1975a, p. 10) in order to increase the "likelihood that multiple frames on a situation will be generated within individual panelists" (Bolger & Wright, 2011, p. 15101). As this study is embedded in the field of education, it can further be argued that it is reasonable to include experts who are within their context of influence potentially able to disseminate and apply the results in practice (Häder, 2000, p. 18).

This approach of choosing stakeholders relates to a procedure from the social sciences in the context of qualitative research known as theoretical sampling.

In order to involve participants who are most appropriate for enhancing the re-search progress, this approach pursues a selection of participants based on theo-retical considerations according to criteria in view of the research aims (Glaser & Strauss, 1967).

With regard to the curricular frame of this study, the participants of the Ber-lin Curricular Delphi Study in Science are also selected according to criteria derived from curriculum theory (cf. 2.3) as recommended for the curricular ap-plication of the Delphi technique (Frey, 1980a, p. 32). Following Mayer (1992, pp. 94-95) and Frey (1980a, p. 32), this study applies four curriculum related criteria for selecting the participants:

- Qualification in the frame of central pedagogical concepts
- Qualification with respect to discourse within the given issue
- Expertise in the area of the curricular issues and in the context of general education
- Different types of affectedness in the curricular process of the given issue

On the basis of the considerations about the sample selection within this study, great emphasis is placed on assembling the sample in such a way that the differ-ent members of the public who are directly or indirectly involved in science education are included in a balanced way. Moreover, the approach of this study, which considers science from an integrated perspective (see 2.1.4) requires a balanced distribution of the participants among the different scientific fields and subjects in order to avoid one-sidedness or domination of one of the science disciplines.

Following Bolte (2003b, p. 12), relevant stakeholders in science education (See 2.3.2) that fulfill the discussed criteria are students, teachers, science educa-tion researchers, scientists, and education policy. An overview of the intended sample of this study is provided in Table 1. The sample of this study is structured according to these major groups on the basis of the notion that the responses of the participants are influenced by their professional background and field of activity (Häder, 2009, p. 107).

Due to previous experiences by Bolte (2003b, p. 11), it is expected that not enough stakeholders who can be assigned to the group of education policy will take part in this study to constitute a sufficiently solid sub-sample with respect to differentiated analyses. Therefore, this study, as previous studies, is likely to proceed without the group education policy.

The group of students included in this study refers to students between the age of 15 and 19, when they attend (or are about to enter) higher secondary level or leave school to find a job or start a vocational trainee program. They have thus experienced different science subjects as basic or advanced courses within their education. The group of teachers is characterized by different stakeholders who are involved in the teaching of science and deal with science teaching respectively. These are represented both by in-service teachers and pre-service teachers (teacher students at university and trainee teachers) as well as by teacher supervisors (such as trainee teacher educators) with different science subjects. In Germany, science education takes place at secondary schools mostly within the traditional subjects of biology, chemistry and physics and is taught from an integrated perspective only at primary level (cf. 2.1.4). Therefore, the science education students taking part in the Berlin Curricular Delphi Study in Science are enrolled in an elementary or secondary level teaching program with the subjects of either biology, chemistry, physics (or a combination of those) or elementary science.

The group of science education researches in the targeted sample of this study refers to educators in the field of science education research and university teaching as well as to science educators who participate or hold posts in didactic associations.

The group of scientists consists of professionals who work in the field of biology, chemistry or physics or in affiliated areas (such as biochemistry, medicine, pharmacy, engineering etc.) at universities, at research institutes, in science associations, or in companies.

The group of education administration refers to science curriculum developers, policy makers, local representatives, personnel in Ministries of Education, and further formal national educational authorities.

With respect to sample sizes in Delphi studies, researchers have made different experiences with the scope of Delphi panels and recommend samples between a total of 10 and 50 panelists (Häder, 2009, p. 96). Yet, with the appropriate sample size being closely connected to the context and specific purpose of a given Delphi study (Ammon, 2009, p. 46), these recommendations mostly relate to Delphi study designs with particular emphasis on inducing group communication processes that lead to establishing consensus among a certain number of experts. In contrast, this study aims at collecting a wide range of views and opinions of selected parts of society, and follows the type of Delphi study that is characterized by primarily attempting to empirically identify and display a group of experts' views on a diffuse issue (cf. Häder, 2009, pp. 29-35). For this type of Delphi study, it is recommended to involve as many participants as possible, as the more experts are included, the more meaningful the results of the study will be (Häder, 2009, pp. 107-110). Also, it is suggested that in general, the more complex the topic of a given Delphi study is, the larger the size of the sample should be (Ammon, 2009, p. 465). With respect to literature recommendations on minimum sample sizes in Delphi (cf. 2.3.3), the intention of allowing for differentiated statistical analysis implies that the stakeholder sub-samples of this study should not fall below a level of at least seven participants after dropout (cf. Becker, 1974, p. 12; Dalkey, Brown, & Cochran, 1969, p. 6).

Table 1
Theoretical Sample Structure of the Study – Groups and Characterization

Sub-Sample	Groups	Areas covered	Intended final sample size
Students	Basic science course Advanced science course	Biology Chemistry Physics	N = 25
Science Teachers	Science education students at university Trainee science teachers Science teachers Trainee science teacher educators	Biology Chemistry Physics Elementary science	N = 25
Science Education Researchers	n.a.	Biology Chemistry Physics Elementary science	N = 25
Scientists	n.a.	Biology Chemistry Physics Others (medicine, pharmacy, engineering etc.)	N = 25
Education Policy	n.a.	Spokespersons of Education Policy in Ministries of Education Formal Educational authorities	____

Based on these concerns, this study intends to involve a minimum number of 25 participants for each of the above defined sub-sample groups, which amounts (disregarding the group of education policy) to a total of at least 100 stakeholders (see Table 1). With regard to anticipated dropout throughout the rounds (cf. 2.3.3.1.8) and the rule of involving a fixed group of participants in a Delphi study (cf. 2.3.3.1.3), the first round targets a number of 30 to 40 participants per sub-sample.

3.2 Instruments for Data Collection and Methods of Data Analysis

The description of the questionnaire designs and of the methods of data analysis is structured according to the three rounds of this study. All statistical preparation and analysis of the data obtained in this investigation is conducted via SPSS. Statistical data analysis procedures chosen appropriate with respect to the questionnaires and research questions include descriptive statistics, identification of central tendencies, multivariate data analysis, and statistical tests with respect to the research hypotheses.

3.2.1 First Round

3.2.1.1 Leading Questions

The first round serves the purpose of accessing the views of different stakeholders about desirable aspects of scientific literacy oriented science education. The leading questions of the first round are:

1. What situations and motives with respect to facilitating science-related educational processes are mentioned?
2. What science related contents, methods and themes are addressed?
3. What competences and attitudes with regard to supporting students in becoming scientifically literate are named?
4. What emphases can be identified in the participants' opinions?

3.2.1.2 Design of the Questionnaire

For of the purpose of the first round, the participants' opinions are initially collected through an open response format (Rost, 2004, p. 59). The response is formulated individually by the participant and has to be coded by the researcher into categories (see 3.2.1.3). The first round questionnaire applied in this study (see appendix) follows Häußler et al. (1980) and Bolte (2003b). The decision to use an open response format in the first round is based on recommendations in Delphi research (Häder, 2009, pp. 87-116). In general, it is assumed that with broad, open-ended questions, a wider range of responses can be gained than from a narrow set of questions – hence, the issue of a given Delphi study is usually approached in a qualitative questionnaire design in the first round. The questions in such qualitative first round should be formulated on a more general level in order to provide the participants with a wide frame for their considerations, as in this way, it can be avoided that the participants feel constrained by guidelines (cf. 2.3.3.1.1).

Based on these considerations, the central question of the Berlin Curricular Delphi Study in Science is formulated in the following way:

Which aspects of science education do you consider meaningful and pedagogically desirable for the scientifically literate individual in the society of today and in the near future?

In the context of the curricular application of the Delphi method, the participants' views and opinions are collected within a certain curricular structure. This structure can be defined by different curricular dimensions such as contexts and situations, topics and concepts, and competences and attitudes (cf. 2.2.2). On the basis of this approach, the overarching question of this study is specified within three open questions in the first round questionnaire, as already proven useful by Häußler et al. (1980) and Bolte (2003b). Following the curricular Delphi method, such specification in terms of a formalized question and answer format systematizes the interrogation, ensures that the statements of the participating experts relate to potential curricular situations, and avoids in this way general statements (Frey, 1980a, p. 32). Moreover, this procedure can be related to insights from research literature that a concise specification of the questions might help to increase potential experts' willingness to participate (cf. 2.3.3.1.1). The specified questions are formulated as follows:

1. *Which situations and motives can be taken as a basis and in which context should science lessons be put in order to facilitate science-related educational processes?*
2. *Which contents, methods and themes related to science should be taught in science lessons?*
3. *Which competences and attitudes should be developed and enhanced to support students in becoming scientifically educated?*

The classical Delphi method is for curricular adaption supplemented by framing the overarching question at the beginning of the interrogation with additional advice and specifying aspects to induce reflection and at the same time avoid stereotype answers Frey (1980a, p. 32). For this reason, several supporting elements are added to the questionnaire in the first round of this study. These elements include an outline of the purpose, context and scope of the study, and an explanatory note with advice for filling in dealing with the questionnaire of the first round (see appendix). Additionally, in order to unify the response format, the participants are asked to formulate their statements in such a way that every answer contains three formal statement elements according to the three questions. An example of a statement according to the three statement elements is provided for clarification and guidance at the beginning of the questionnaire. To provide the participants with sufficient response opportunity, the round one questionnaire includes ten identical form sheets, each with the aforementioned predefined question and response format according to the three specified questions.

3.2.1.3   Data Analysis

The participants' individually formulated statements are in the course of the qualitative and quantitative analyses organized, labeled, and summarized into categories (statement bundles).

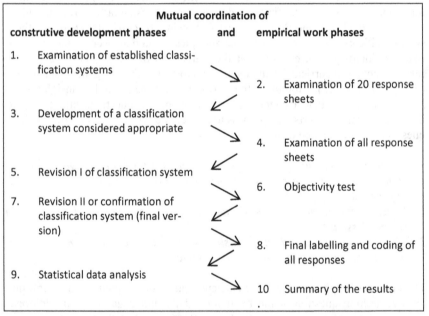

Figure 2. Procedure of analyzing the statements in round 1 following Bolte (2003b)

In this way, a consolidation of the statements of the first round into meaningful categories and a quantification of the data takes place. On this basis, the data is statistically evaluated via SPSS by determining category frequencies as well as testing for differences in the distribution of the category frequencies depending on different groups. The general procedure of the qualitative and quantitative steps of analyzing the participants' statements follows Bolte (2003b) and is shown in Figure 2.

3.2.1.3.1 Qualitative Analysis

In the following part, the procedure of the qualitative analysis of the open-text answers received from the participants in the first round of this study is described. The overall aim of the qualitative analysis is a systematization of the participants' statements in order to display their responses and allow for quantitative analyses

First, as shown in Figure 2, classification systems[14] of previous curricular Delphi studies on science subjects (Bolte, 2003b; Häußler et al., 1980) as well as other systematizations of curricular dimensions of science education from literature (e.g. AAAS, 2001; KMK, 2005a, 2005b, 2005c; MNU, 2003; OECD, 2007b; Schaefer, 2010a, 2010b) are examined (step 1). In this way, the initial approach to investigating the participants' statements is based on aspects from major fields of physics, chemistry, biology, and earth and space science that are considered to be of relevance to real-life situations. On this basis, the statements of 20 questionnaires are analyzed, structured, and systematized (step 2) by means of a software application for qualitative and mixed methods data analysis (MAXQDA). In the course of this analysis, the statements of the participants are transferred to a classification system on the basis of which the data can be described more compactly though categories. For this purpose, the statements are grouped and summarized into categories[15], following the principles of qualitative content analysis (Gläser & Grit, 2010; Krüger & Riemeier, 2014; H. O. Mayer, 2006; Mayring, 1983; Mayring & Gläser-Zikuda, 2008). The qualitative content analysis approach (Mayring, 1989, 1993) is an approach to structured and intersubjectively verifiable interpretation of extensive written material. In contrast to a global analysis, qualitative content analysis involves a more precise analysis and aims at establishing an elaborated category system, which provides the basis for a comprehensive interpretation of the material. In such structuring analysis, the developed category system is refined for the final analysis after a test trial. In the context of content analysis, categories function as variables or attributes of variables. As one category is usually not sufficient, text interpretations generally

---

14   The term "classification system" denotes in this case a set of categories to which statements can be assigned.
15   The term "category" is throughout this study used to refer to a group of interrelated statements in the sense of a "statement bundle", which is thus not to be understood as exclusively defining.

operate on the basis of a system of categories (Mayring, 1989, 1993). For this reason, the term category system is also referred to as classification system or interpretation scheme. For categories that occur very frequently and appear very broad, subcategories can be formed.

Ideally, category systems are either inductively derived from the material or deductively (theory-based) used to approach the material. In the inductive method, a separate category system is developed for every participant and thus characterizes the individual case. In deductive approaches, a category system applied to all text material represents a frame which describes the different cases through their individual category occupations. In practice, mixed forms are common, in which an a priori established category system is extended and refined upon review of the material. A category system can either operationalize constructs for the purpose of hypothesis testing or pose open questions or topics for the purpose of hypothesis search and description. The coding takes places on the basis of the category system and means assigning text parts (in this case: statements or parts of statements) to categories. The quality of coding depends largely on the definition of the categories. Only if the categories are well defined and exemplified in written coding instruction, the material can be processed accurately. On the basis of the coded individual cases, inter-subjective comparisons are possible. The case of all categories being occupied by a minimum number of text examples is referred to as a saturated category system. Empty or nearly empty (unsaturated) categories suggest that the respective constructs might be irrelevant for the study, poorly defined, or that not enough cases were examined. If hypotheses are formulated prior to the study, the frequency information in the category system forms the basis for hypothesis testing (Bortz & Döring, 2006, pp. 329–331). In order to take account of both the great diversity of the responses and the necessity for condensing the wide range of information, the statements are summarized as much as possible while staying as differentiated as necessary.

After this first consideration of statements, an initial classification system for the analysis of the participants' statements is established (step 3). As implied by the three-part format of the questionnaire, the classification system is subdivided into three different parts (I: situations, contexts, and motives, II: fields, and III: qualifications). A recommended further subdivision of part II (field) into part IIa (scientific concepts and topics) and part IIb (scientific disciplines and fields) according to Bolte (2003b, p. 13) is adapted as well. On the basis of this classification system, all questionnaires are categorized in a process of open coding (step 4). In this process, statements are either assigned to already existing categories, or additional categories are developed. In addition, if applicable, similar categories are summarized and subsumed under a broader category. In this way, the initial category system is revised, modified, and supplemented by further categories (step 5).

As qualitative evaluation methods interpret verbal, non-numerical material and proceed in inter-subjectively verifiable steps, valid interpretations must reach a consensus among the coders (Bortz & Döring, 2006, p. 331). Hence, after this revision, a sample of 20 questionnaires is coded by two independent raters to determine the inter-rater agreement (step 6). The inter-rater agreement is used in empirical research to refer to the degree of consensus or homogeneity in independent coding by different raters. In other words, it provides a measure of the extent to which these raters assign open-text answers to the same categories. As a result, the inter-rater agreement provides insights into the extent to which the results of coding are independent from the rater. For this reason, inter-rater agreement as a subclass of evaluation objectivity is also referred to as coding objectivity (Rost, 2004, p. 79). Objectivity is an essential quality property in empirical research which generally refers to interpersonal consensus in terms of two independent researchers reaching comparable results within the same investigation and within application of the same methods. This ensures inter-subjective replicability and thus interpretability of results independent from the rater (Bortz & Döring, 2006, p. 326). Hence, through determining the inter-rater agreement at this point, the qualitative analysis of the statements in terms of coding the participants' statements into categories is tested for objectivity. The inter-rater agreement (q) in this study is determined through a formula following Bolte (2003b, p. 15), Häußler et al. (1980, p. 115) and Mayer (1992, p. 61):

$$q = \frac{2N_+}{2N_+ + N_-}$$

N$_+$: Number of cases in which positive category coding of both raters matches
N$_-$: Number of cases in which only one rater assigns a category

In case of complete congruence of the coding, q equals 1. It should be noted that this quotient only takes into account positive coding of a category. In this way, it does not consider the number of matching negative category coding, that is, how often the raters agree about a certain category being not addressed in a given statement. Hence, this quotient is a rather strict measure. If the outcome of the inter-rater agreement indicates it as being necessary, the classification system is revised again and the previous steps are repeated. If results of the objectivity test are sufficient, the established classification system is confirmed (step 7) and maintained for final labeling and coding of all statements (step 8).

3.2.1.3.2   Preparation of the Data for Quantitative Analysis

In order to prepare the results of the qualitative analysis for quantitative statistical analyses, the data is entered in SPSS on the basis of the developed classification system. The responses are coded on a dichotomous level (Rost, 2004, p. 83):
If a category from the applied classification system is mentioned on a form sheet,
the corresponding category is coded with "1", and every category not mentioned
is coded with "0". Due to the dichotomous coding, every mentioned category is
only counted once per form sheet, although a category can be addressed several
times within the same form sheet. Due the open-text response approach of the
first round, it is possible that one statement references multiple categories.

3.2.1.3.3   Descriptive Statistics

In order to gain a more differentiated overview of the empirical data, descriptive
statistical analyses are carried out, taking into account both the total sample and
the stakeholder sub-samples (step 9 of the procedure of analyzing the participants' statements of the first round, see Figure 2).With the main purposes of the
quantitative analysis to gain insights about the response behavior of the participants, identify what general statements and emphases can be retrieved from the
participants' responses, and determine which distinctive features appear in a
differentiated analysis of the sub-samples, the analyses are carried out by taking
particularly into account the following characteristic values:

- Total number of form sheets filled out by the participants
- Average number of form sheets per person
- Total number of categories mentioned by the participants
- Average number of categories mentioned per person
- Relative frequencies of the categories regarding both the total sample and the
  sub-samples

Due to the dichotomous coding, the mean value of a category represents its relative frequency with respect to all responses. As referring to different categories
within the same form sheet is possible, the addition of the individual relative
category frequencies can result in percentages higher than 100 percent. When
determining the relative frequencies of the categories, multiple references to a
category by the same person are treated as single category entries to avoid an
imbalance due to personal emphases of the participants. This procedure is necessary to account for the participants' opportunity to use the given number of ten
form sheets in a flexible way. Special emphasis is placed on identifying catego-

ries mentioned rarely (n ≤ 5%) or very often (n ≥ 25%). These analyses are carried out both for the total sample and for the different sub-samples. As a result of this analysis, categories mentioned by many stakeholders as well as by few stakeholders, both within the total sample and the different sub-samples, can be identified. Also, this procedure provides information about the degree of differentiation in which the participants provide render their statements, and first insights into agreement and disagreement between the different sub-samples.

3.2.1.3.4 Statistical Testing Procedures

Statistical testing procedures are used to address difference hypotheses concerning the investigated data. In general, statistical testing procedures determine the probability of the results being based on random distribution (Bortz, 2005, p. 135). As the results of quantitative content analysis consist of count data (Bortz & Döring, 2006, p. 149), procedures of hypothesis testing can be applied to the data resulting from the first round.

The quantitative data of the first round allows for investigating the distribution of category frequencies. Such investigation is essential with respect to testing the difference hypotheses regarding the opinions of the four sub-samples (students, science teachers, science education researchers, and scientists, cf. hypothesis 5b), as well as a comparison of the responses of the students and the group of the adult sub-samples on aspects of desirable science education (cf. hypothesis 5c),

For the purpose of analyzing frequency differences of certain features, as intended in these comparisons, nominal data procedures are required (Bortz, 2005, p. 154). An appropriate nominal data procedure for investigating frequency distributions that result from a classification of testing objects into different groups is provided by the chi-square test. The chi-square test refers to a group of statistical testing procedures with the sampling distribution of the test statistic equaling or asymptotically resembling a chi-square distribution when the null hypothesis is true. The null hypothesis represents an assumption about a theoretically expected distribution of frequencies. The expected frequencies according to the null hypothesis are compared to the empirically observed frequencies on the basis of the chi-square value. If the empirical chi-square value is sufficiently improbable, the null hypothesis can be rejected and the alternative hypothesis assumed. In this way, the chi-square test can be used to determine whether there is a significant difference between expected and observed frequencies in one or more categories, or if the differences are based on sampling variation (Rasch, Friese, Hofmann, & Naumann, 2006, p. 172).

An application of chi-square testing procedures presumes that the different observations are independent from each other, that the test persons can unambiguously be assigned to a classification or attribute combination, and that the expected frequencies are in 80% of the cells characterized by N≥5 (Rasch et al., 2006, p. 198). These requirements are fulfilled by the investigated data of the first round.

Comparison of the Sub-Samples

With the aim of testing for differences among the four sub-samples (hypothesis 5b), a two-dimensional chi-square test for independence (also referred to as contingency analysis) is used, as this comparison includes several categorical attributes with at least two gradations. The test persons are assigned in this case to the possible combinations of the gradations on the basis of a cross tabulation (contingency table). In this way, the two-dimensional chi-square procedure tests the investigated attributes for independence (Rasch et al., 2006, p. 185).

Comparison of the Group of Students and the Group of Adult Sub-Samples

Applying the difference hypothesis 5c to the data of the first round implies that the category frequencies in the group of students and in the group of adult sub-samples differ from each other. Featuring thus two categorical attributes with two gradations in each case, both attributes are dichotomous. Hence, a 2x2 chi-square test in a cross tabulation design (contingency table) as a particular sub-type of two-dimensional chi-square tests for independence has to be applied. Based on an assumption about the theoretically expected distribution, the expected distribution of given cells can in this test be determined and compared with the observed values. This test is particularly used in case of two dichotomous attributes and enables propositions as to whether the two observed features are stochastically related in any form, with the null hypothesis of the test postulating stochastic independence of the two features and the alternative hypothesis claiming a certain correlation between the levels of one feature and the levels of the other (Rasch et al., 2006, p. 196).

3.2.2   Second Round

3.2.2.1   Leading Questions

The general purpose of the second round is to gain more solid insights into opinions on desirable science education and to represent a synopsis of the responses collected in the first round. Hence, the leading questions of the second round are:

1. What priorities regarding aspects of desirable science education can be identified in the participants' assessments?
2. What degree of realization in practice regarding aspects of desirable science education can be identified in the participants' assessments?
3. What priority-practice differences regarding aspects of desirable science education can be identified in the participants' assessments?
4. What emphases can be identified in the participants' assessments?
5. What concepts of desirable science education can be identified in the category combinations provided assembled by the participants?

3.2.2.2   Design of the Questionnaire

With respect to the purpose of the second round, the stakeholders are provided with an opportunity to reconsider and, if applicable, revise their views and emphases expressed in the first round in light of the general opinion of the expert panel as well as to further process the given set of categories. Therefore, the questionnaire of the second round is generated on the basis of the results of the first round. In this way, the findings from the first round are elaborated and specified in the second round.

More specifically, the aim of the second round is to determine aspects that are considered most and least important for science education as well as gain insights into the extent to which these aspects are present in current science education. This yields insights into areas in which priority and current realization of desirable aspects of science education drift apart in the opinions of the participants. In addition, the second round intends to identify empirically based concepts of desirable contemporary science education. For this purpose, all participants from the first round of the Berlin Curricular Delphi Study in Science are in the second round presented with a two-part questionnaire (see appendix) and asked to assess and further process the provided categories.

Part I of the Questionnaire

Part I of the questionnaire contains a list with all categories established in the first round analyses of the Berlin Curricular Delphi Study in Science. For clarity reasons, the list maintains the three-part format of the first round questionnaire (cf. 3.2.1.2), which is divided into contexts, situations, and motives, concepts and topics, and competences and attitudes. In order to avoid misinterpretation of the category labels, a glossary with a description and explanation of the categories and examples is provided (see appendix). This glossary is embedded in the online questionnaire and enclosed to the paper-and-pencil questionnaire.

The first part of the questionnaire requires weighted assessment of these categories from two different perspectives. In order to identify aspects that are considered most and least important for science education, the participants are asked to prioritize the given categories. To classify aspects perceived as very often and rarely present in current science education, the participants are as well as asked to assess to what extent the aspects expressed by the categories are realized in science education practice.

The participants' assessment of the categories is guided by the following questions:

1. Which priority should the respective aspects have in science education? (Priority dimension)
2. To what extent are the respective aspects realized in current science education? (Practice dimension)

A six-tier rating scale is applied for collecting both assessments, as shown in the following example (Figure 3). As a middle neutral response option is not available in even-point scales, this six-tier scale is designed in terms of a "forced choice" structure. The decision for using an even-point scale is based on the notion that a middle response option has been found to entail the risk of ambiguous meaning, as participants might choose it if they are unsure, refuses a response, or does not consider the item as suitable (Rost, 2004, p. 67). With the level of measurement determining which mathematical operations are allowed and what interpretations are possible, the question regarding the type of scaling of the data is of crucial importance with respect to the data analysis (Bortz, 2005, p. 18). Formally, part I of the second round questionnaire can be characterized as a rating scale, since it relates to bipolar, non-dichotomous data composed of a range of values that measure opinion. As the participants are in this part of the questionnaire given the opportunity to specify the level of their personal estimations on different predefined gradations, its scheme can also be compared to a

Likert-type scale design, which is characterized by a bipolar scaling method that measures positive or negative assessment of a given statement and thus represents a specific type of rating scale (Rost, 2004, p. 50). Rating formats are a frequently used subgroup of bound response formats, which are, in contrast to open response formats, characterized by offering a choice of response options.

Part I: Situations, contexts and motives	Which **priority** should the respective aspects have in science education?	*To what **extent** are the respective aspects realized in current science education?*
Please assess the following categories according to the two questions stated.	1 = very low priority 2 = low priority 3 = rather low priority 4 = rather high priority 5 = high priority 6 = very high priority	*1 = to a very low extent* *2 = to a low extent* *3 = to a rather low extent* *4 = to a rather high extent* *5 = to a high extent* *6 = to a very high extent*
Category 1	[1] [2] [3] [4] [5] [6]	*[1] [2] [3] [4] [5] [6]*
Category 2	[1] [2] [3] [4] [5] [6]	*[1] [2] [3] [4] [5] [6]*
...	[1] [2] [3] [4] [5] [6]	*[1] [2] [3] [4] [5] [6]*

Figure 3. Example of part I of the questionnaire in the second round – assessment of categories

In general, rating scales are distinguished by multi-tiered response categories that symbolize marked gradations of a continuum and are thus assumed to represent a certain rank order to the respondents. These response categories are structured in a unipolar or bipolar way and can be represented on a verbal, numeric, or symbolic level. In contrast to dichotomous response formats, rating scales enable participants to express their responses on a more differentiated level and are thus more sophisticated in terms of informative content. In addition, response categories of rating scales are formulated on an item-unspecific level, which means that the same assessment scheme applies to several or all response categories of a questionnaire. As in general, it is not assumed that the response options on rating scales are perceived as equidistant by the respondents, the data of rating scales is usually represented through an ordinal level of measurement (Rost 2004, pp. 64-68). The ordinal scaling type is characterized by allowing for rank order between different values, but no claims can be made about the degree of difference between the data. For this reason, the results within ordinal scaling can be described through the median as a statistical parameter for indicating central tendency but not through the arithmetic mean. The arithmetic mean as a measure of central tendency, along with measures of statistical dispersion such as range and standard deviation, requires at least interval scaling, as equal distance assumption is an essential prerequisite for such statistical procedures and only metric scales allow for intervals between different values (Bortz, 2005, pp. 19-21).

However, the distinction between ordinal and interval scaling is often ambiguous in practice. In order to allow for more sophisticated statistical operations, conventional research practice usually assumes that the levels of the rating scale represent an interval scale (Bortz & Döring, 2006, pp. 176-181). For this reason, interval scaling is frequently applied in case of rating scales (Bortz, 2005, p. 26). For an approximation to interval scaling, numbers are often assigned to items in rating scales (Rost, 2004, p. 67). This practice is based on the assumption that the conformation of a research hypothesis is rather impeded than facilitated through an assumption of a wrong scaling type (Bortz, 2005, p. 26). Also, such scales are found to perform similarly to scales perceived as equidistant (Labovitz, 1967, p. 160). Therefore, it is argued that under certain circumstances assigning numbers to ordinal categories is both legitimate and useful (Labovitz, 1967, p. 153).

In this study, the applied six-tier scale is assumed to represent a rather large rating scale so that the participants perceive the scale mainly as a continuum and thus interpret the different scale points as equidistant. This assumption is fostered by various aspects on the construction of the response scale: As with verbal labelling of the scale points, the meaning of the different response levels is inter-subjectively unified, and assigning numbers promotes an understanding of the scale in terms of equally distant intervals, a combination of both devices is applied. Moreover, the verbal response options in the questionnaire are formulated in a symmetric scheme. In addition, the corresponding visual scale on the questionnaire is displayed as a horizontal sequence of boxes on which the participants mark their responses (Rost, 2004, pp. 67-68).

As a result, the assumption of the participants perceiving the scale points of the given rating scale as equidistant allows for treating the data of this questionnaire part as compliant to interval scales and thus interpreting the data on a metric level. For this reason, interval scaling is assumed with respect to the statistical operations of the data gained in this first part of the questionnaire (see 3.2.2.3).

Part II of the Questionnaire

As scientific literacy is a complex construct (Gräber & Bolte, 1997; Gräber & Nentwig, 2002), the enhancement of scientific literacy is not possible by referring to the different aspects in isolation. Instead, the promotion of scientific literacy is only possible if the complexity of the scientific literacy construct is considered more comprehensively. Therefore, the empirically identified aspects from the first round need to be considered in relation to conceptual approaches as well. For this purpose, the participants are in part II of the second round asked to combine categories from the given set of categories that seem meaningful to them in combination (see appendix).

In order to avoid overgeneralization as well as imbalances in the grouping of the categories, a combination should include at least one category of each column up to a limit of five categories per column. The columns in the second part of the questionnaire represent overarching category sections on the basis of the three-part format of the first round questionnaire (cf. 3.2.1.2), which is divided into contexts and motives, concepts and topics, and competences and attitudes. To provide the participants with sufficient opportunity for compiling combinations, part II of the second round questionnaire includes ten identical form sheets, each containing the set of categories emerged from the first round and structured according to the previously described format. In this way, the participants' opinions are more condensed in the second part of the second round

3.2.2.3 Data Analysis

3.2.2.3.1 Descriptive Statistics

The importance and extent of realization of the given categories in the science classroom is evaluated with descriptive statistical approaches to the stakeholders' priority and practice assessments. In order to gain more detailed insights into aspects that are considered more and less important for science education as well as to determine the extent to which these aspects are present in current science education, the central tendency in the assessments as represented through mean values is considered. For this purpose, the mean values of the priority and practice assessments are calculated for each category.

In addition, to identify areas in which priority and practice drift apart and hence need improvement, as well as to identify those areas that are more adequately represented in the opinions of the participants, priority-practice differences (PPD) are identified for every category. These differences are calculated for every category by subtracting the practice value from the priority value: $\Delta PPD = x_{Priority} - x_{Practice}$. For all priority-practice differences, mean values are determined as well. In case of a positive difference, the participants consider the corresponding category to be less present in the classroom than its importance requires. Large positive differences indicate categories considered as very important but only rarely realized, pointing to areas of science education that are in particular need for action. In case of small differences, the realization of a category in practice approximately corresponds to its given importance. If the difference is negative, the corresponding category is considered to be more realized in current science education than its importance implies. In order to account for both the general opinion and differentiated analyses, mean values of priority, practice, and priority-practice differences are determined for the total sample as well as for the four different sub-samples.

As sub-samples can have different response rates in Delphi studies, the total sample might be imbalanced with respect to the sub-sample sizes. Moreover, as no new participants are added after the first round due to the principle of involving a fixed group of participants in Delphi studies, dropout of participants after the first and second round can lead to particular over- and under-representation of a sub-sample group so that the total sample shows unequal distribution of the sub-sample sizes. As a result, average assessments of the total sample can be disproportionally influenced by the assessments of the sub-samples and do thus not display the views of an equally composed sample. For this reason, the afore-mentioned calculations for the total sample are reconducted with weighted data to transfer the different sub-samples' assessments in equal proportions into the representation of the total sample's assessments. This procedure allows for a more accurate representation of the assessments of the total sample in terms of comparability and further interpretation. In this way, normalization of the data takes place, which represents a further quality feature of empirical studies (Rost, 2004, p. 41) and allows in this way to convey a clearer picture of the total sample's assessments with respect to comparability and further interpretation.

3.2.2.3.2 Statistical Testing Procedures

The difference hypotheses address a comparison of the views of the four sub-samples (students, science teachers, science education researchers, and scientists; hypothesis 5b) and a comparison of the views of the students and the group of the adult sub-samples (hypothesis 5c). On the basis of the quantitative data gained throughout the second round, statistical testing procedures regarding these difference hypotheses are applied with respect to the assessments of the categories representing aspects of desirable science education.

When choosing adequate statistical testing procedures for the quantitative data which emerged from the second round, a number of criteria have to be considered. In general, parametric and non-parametric testing procedures can be distinguished. The main difference between parametric and non-parametric tests relates to demands regarding the distribution of values in the total population. While parametric tests require normal distribution of the data that is tested, this does not apply to non-parametric tests. For smaller samples, it is in case of doubt recommended to use non-parametric tests instead of parametric tests, as normal distribution can only be assumed for samples with N>30 (Bortz & Döring, 2006, p. 94). Moreover, the choice of appropriate testing procedures depends on the type of scaling of the data. In general, non-parametric tests have lower demands on the level of scaling than parametric tests. As non-parametric tests are based on rank values (comparing median values), ordinal scaling is sufficient, while parametric testing procedures usually compare arithmetic mean values and therefore require at least interval scaling of the data. Hence, in case of data with lower levels of scaling, non-parametric tests have to be applied (Bortz, 2005, p. 136).

As a result of these considerations, non-parametric testing procedures are used for the purpose of comparing the priority and practice assessments as well as the priority-practice differences of the different sub-samples. This choice is in particular based on the fact that although the data is treated on an interval scaling level, the sub-sample sizes in the second round with a final targeted number of 25 participants per sub-sample are likely to fall below a number of N=30 and thus normal distribution cannot be assumed for the data of the sub-samples that are to be compared.

Comparison of Sub-Samples

For a comparison of the priority and practice assessments as well as the priority-practice differences of the categories in the different sub-samples, significance tests for independent samples have to be used. With respect to the difference hypotheses (cf. hypotheses 5b and 5c), pairwise comparisons of the sub-samples' assessments are taken into account. An appropriate procedure can be found with the Mann-Whitney-U test. As a non-parametric testing procedure for independent samples, this test investigates whether two different samples differ in terms of their central tendency, which is determined on the basis of median values. Statistically significant outcomes of this test indicate that differences between the data of the compared assessments do not result from random distribution (Bortz, 2005, pp. 150-153). In this way, statistically significant differences between the assessments of the different investigated groups are identified.

As the comparison between the assessments of the students and the group of adult sub-samples can be considered a particular case of pairwise comparison of parts of the total sample, the same conditions apply as for a comparison of the four sub-samples, and the Mann-Whitney-U test is applied in this case as well. This testing procedure is in these comparisons carried out for each category.

Statistically significant outcomes indicate that differences between the data of the compared assessments do not result from random distribution (Bortz, 2005, pp. 150-153). In this test, a significance level (p) of 0.05 is applied, which implies that in case of the null hypothesis being true, the probability of the results of a study leading to a rejection of the null hypothesis in favor of the alternative hypothesis does not surpass a threshold of 5% (Bortz, 2005, p. 114).

3.2.2.3.3 Hierarchical Cluster Analysis

Part II of the second round questionnaire yields groups of categories which the participants consider meaningful for desirable contemporary science education in combination. Providing in this way quantitative data, multivariate methods such as cluster analysis can be used for structuring and summarizing the participants' views. In order to identify empirically based concepts of desirable science education, these combinations are processed through hierarchical cluster analysis. Hierarchical clustering as a type of data mining represents a systematic distance-based method for identifying similarity structures (patterns) in a data set. The overall goal of such procedures is to extract information from a given data set and convert it into a coherent structure for further use. Clusters consist of values that are characterized by a closer distance (similarity) between each other than to values in other clusters. In general, two major types of clustering can be distinguished. In the divisive clustering method, also referred to as a "top down" approach, all objects are initially considered within one cluster, which is recursively subdivided into smaller clusters along the hierarchy until each cluster consists of only one object. In contrast, the agglomerative clustering method starts at the bottom of the hierarchy with each object representing one individual cluster. In the course of this procedure, pairs of clusters are gradually merged into larger clusters until all objects belong to one cluster. This process is also known as a "bottom up" approach. The results are usually visualized in a dendrogram, a tree diagram frequently used for illustrating the structure of the clusters produced by hierarchical clustering (Bortz, 2005, pp. 571-573).

With the purpose of identifying concepts from established clusters as aimed at in part II of the second part, the "bottom up" approach is chosen. When merging a pair of clusters in agglomerative clustering, different criteria are available to determine which clusters are to be fused (Bortz, 2005, p. 572). The criterion applied to the cluster analysis of this study is characterized by the Ward method with the squared Euclidean distance as interval scale related distance measure that is calculated between all objects. The Ward method is an agglomerative clustering method that aims to create clusters that are as homogeneous as possible by merging preferably those objects that facilitate the dispersion in a group at the lowest level. Thus, in the first steps of merging according to the Ward method, clusters are preferably formed in regions with high object density. With advancing fusion, the Ward algorithm tends to balance differences in the occupation numbers of the clusters and facilitates in this way that clusters with similar sizes are gradually established (Bortz, 2005, pp. 575-578). In order to avoid mini-clusters that are too differentiated for interpretation as well as clusters too broad and thus not interpretable on a content level with regard to identifying concepts of science education, a reasonable level with regard to the number of clusters has to be chosen (Häußler et al., 1980, p. 213).

3.2.3 Third Round

3.2.3.1 Leading Questions

With the aim of gaining more detailed insights into the participants' opinions regarding concepts of desirable science education which are derived from the cluster analysis of the category combinations provided by the participants in the second round, a final third round is designed on the basis of these considerations, the leading questions of the third round are:

1. What priorities regarding concepts of desirable science education can be identified in the participants' assessments?
2. What degree of realization regarding concepts of desirable science education can be identified in the participants' assessments?
3. What priority-practice differences regarding concepts of desirable science education can be identified in the participants' assessments?
4. What emphases can be identified in the participants' assessments?
5. How are these assessments differentiated according to educational levels?

3.2.3.2 Design of the Questionnaire

Following the aims of the third round, all participants from the second round of the Berlin Curricular Delphi Study in Science are in the third round presented with a questionnaire that asks them to assess and further elaborate on the concepts which resulted from the responses in the second round (see appendix). In this way, the third round questionnaire builds on the responses gained from the second round. In order to identify the importance of the concepts and to determine their degree of realization, as well as to discover priority-practice differences, the questionnaire inquires weighted assessment of these concepts in terms of priority and practice in the same way as in the second round (cf. 3.2.2.2). The assessment of the concepts is guided by the following questions:

1. Which priority should these concepts have in science education? (Priority dimension)

2. To what extent are the given concepts realized in current science education?
   (Practice dimension)

Analogously to the first part of the second round (cf. 3.2.2.2), the participants'
priority and practice assessments are collected on a six-tier rating scale. First,
these assessments are related to science education in general (Figure 4).

With the aim of further specifying the stakeholders' opinions on concepts of
science education und investigating their views in more detail, participants' assessments are in addition to the general assessment of the concepts' priority and
presence in science education also asked for expectations at different levels of
education. For this purpose, the participants' assessments are in a second step
differentiated according to educational levels. More specifically, the participants
are asked to assess the concepts' priority and degree of realization in practice
with respect to four different educational levels associated with the greater context of general science education: pre-school, elementary level, lower secondary
education, and higher secondary education (Figure 5).

Concepts Please assess the following concepts according to the two questions stated.	Which **priority** should the respective concepts have in science education?	To what ***extent*** are the respective concepts realized in current science education?
	1 = very low priority 2 = low priority 3 = rather low priority 4 = rather high priority 5 = high priority 6 = very high priority	*1 = to a very low extent* *2 = to a low extent* *3 = to a rather low extent* *4 = to a rather high extent* *5 = to a high extent* *6 = to a very high extent*
Concept 1	[1] [2] [3] [4] [5] [6]	[1] [2] [3] [4] [5] [6]
...	[1] [2] [3] [4] [5] [6]	[1] [2] [3] [4] [5] [6]

Figure 4. Example of the first part of the questionnaire in the third round – assessment of concepts of
science education in general

With respect to the European frame in which this study is carried out among
other national Delphi studies (cf. 2.4) and in order to allow for project-wide,
cross-national comparability of results (Gauckler et al., 2014; Schulte & Bolte,
2014b), these levels are based on a differentiation according to the International
Standard Classification of Education (ISCED). The ISCED describes educational
levels as "[a]n ordered set, grouping education programmes in relation to grada-

tions of learning experiences, as well as the knowledge, skills and competencies which each programme is designed to impart", and thus reflecting the "degree of complexity and specialisation of the content of an education programme, from foundational to complex" (UNESCO, 2012, p. 81).

Concepts Please assess the following concepts according to the two questions stated.	Level of Edu-cation	Which **priority** should the respective concepts have in science education? 1 = very low priority 2 = low priority 3 = rather low priority 4 = rather high priority 5 = high priority 6 = very high priority	To what **extent** are the respective concepts realized in current science education? 1 = to a very low extent 2 = to a low extent 3 = to a rather low extent 4 = to a rather high extent 5 = to a high extent 6 = to a very high extent
Concept 1	Pre-school	[1] [2] [3] [4] [5] [6]	[1] [2] [3] [4] [5] [6]
	Elementary level	[1] [2] [3] [4] [5] [6]	[1] [2] [3] [4] [5] [6]
	Lower second-ary education	[1] [2] [3] [4] [5] [6]	[1] [2] [3] [4] [5] [6]
	Higher second-ary education	[1] [2] [3] [4] [5] [6]	[1] [2] [3] [4] [5] [6]
...	...	...	...

Figure 5. Example of the second Part of the Questionnaire in the Third Round – Assessment of the Concepts According to Different Educational Levels

The ISCED distinguishes eight educational levels (UNESCO, 2012, pp. 25–61): Level 0 (early childhood education) includes initial educational development and pre-primary education. It introduces young children to organized instruction to prepare them for entry into primary education (UNESCO, 2012, p. 26). This level is referred to as "pre-school" in the questionnaire. Level 1 (primary educa-tion) describes education that is designed to provide students with fundamental skills and establish a solid foundation for learning and understanding core areas of knowledge and personal development in preparation for lower secondary education. In general, this level focuses on learning at a basic level of complexity with little, if any, specialization and lasts between 4 and 7 years. Usually, stu-dents enter this level between ages 5 and 7 (UNESCO, 2012, p. 30). In the ques-tionnaire, this level is called "elementary level". Level 2 (lower secondary edu-cation) refers to programs that are intended to build on the learning outcomes of primary education with the aim of laying the foundations for lifelong learning and human development, and are usually organized around a more subject-oriented curriculum. Hence, this level also requires teachers with more specialized training. Usually, this level is entered by students between ages 10 and 13 and ends after a duration of 8 to 11 years of education from the start of primary education. In many education systems with compulsory education legislation, the end of lower secondary education coincides with the end of compulsory (general) education (UNESCO, 2012, p. 33). Level 3 (upper secondary education) relates to programs that complete secondary education in preparation for tertiary education or provide skills relevant to employment, or both. It is characterized by more varied, differentiated, specialized and in-depth instruction than level 2. Upper secondary education is entered by students between ages 14 and 16. Generally, the duration of this level ranges between 11 to 13 years of education since the beginning of primary education and ends around age 17 or 18 (UNESCO, 2012, p. 38). Levels 4-8 (UNESCO, 2012, pp. 43-46) are not represented in the questionnaire, as they relate to post-secondary non-tertiary and tertiary education, which are not included in the focus of this study.

3.2.3.3   Data Analysis

The data of the third round is investigated via SPSS through descriptive statistics and statistical testing procedures with respect to the research questions and difference hypotheses applicable to this round. In the following, I will outline the statistical operations and underlying considerations for the analysis of data gained in the third round.

3.2.3.3.1   Descriptive Statistics

The importance and degree of realization of the concepts of desirable science education in current science education is first investigated on the basis of descriptive statistical analysis of the stakeholders' priority and practice assessments. As the questionnaire design of the third round follows the design of the questionnaire in the second round (cf. 3.2.2.2) and thus ordinal scaling can be assumed for the third round questionnaire as well, the central tendency in the assessments can be described by mean values. To gain insights into the priorities and perceived realization of the concepts regarding science education both on general and different educational levels, the mean values of the priority and practice assessments are calculated for each concept, both on the general level and different educational levels. Moreover, to identify concept related areas with an imbalance between priority and degree of realization in practice, as well as to distinguish areas that are according to the participants adequate, priority-practice differences are determined for every concept – again for both the general level and different educational levels. Analogously to the data analysis of the second round, these differences are calculated by subtracting the practice value from the priority value: $\Delta PPD = x_{Priority} - x_{Practice}$. Mean values are calculated for all priority-practice differences. If a positive difference occurs, the corresponding concept is perceived as underrepresented, which means it is realized in practice to a lower degree than its priority would require. Therefore, large positive priority-practice differences point to concepts that are in particular need for improvement. In case of negative differences, the corresponding category is seen as realized to a greater extent in practice than its importance implies and is thus overrepresented. Small priority-practice differences indicate that the perceived degree of realization of a concept mostly corresponds to its estimated importance.

To allow for a representation of the sample's general opinion as well as for differentiated analyses and difference tests, mean values of priority, practice, and priority-practice differences are determined for the total sample as well as for the different sub-samples. Hence, the descriptive statistics include three emphases:

a) General assessment of the three concepts of science education by the total sample
b) Assessment of the three concepts of science education differentiated according to different educational levels by the total sample
c) General assessment of the three concepts of science education by the sub-samples

As mentioned before, differing response rates can occur for sub-samples in Delphi studies, leading to over- and under-representation of certain sub-samples so that the total sample is not represented by equal distribution of the sub-sample. As Delphi studies involve a fixed group of participants and therefore new participants cannot be added after the first round, this effect is likely to increase throughout the rounds. Thus, an imbalanced distribution of the sub-sample sizes within the total sample is likely to result in the third round as well. For this reason, the descriptive statistical analyses for the total sample are in the third round as well reconducted with weighted data.

3.2.3.3.2 Statistical Testing Procedures

With the purpose of continuing in the third round to compare the views of the four sub-samples (students, science teachers, science education researchers, and scientists) and the views of the students and the group of the adult sub-samples on desirable science education (cf. hypotheses 5b and 5c), as well as to analyze whether the concepts differ from each other in their assessments (cf. hypothesis 1c), further statistical testing procedures are applied to the quantitative data gained in the third round. Because of the similarity between the questionnaire design of the second and third round, the same considerations as for the choice of appropriate testing procedures in the second round (cf. 3.2.2.3.2) apply to the selection of testing procedures in the third round.

Comparison of the Sub-Samples

Although the data is treated on an interval scaling level, the sub-sample sizes in the third round with a targeted number of 25 participants per sub-sample fall below a number of $N = 30$ and thus normal distribution cannot be assumed for the data of the sub-samples. Hence, in the same way as in the second round, the Mann-Whitney-U test (cf. 3.2.2.3.2) has to be applied as an appropriate non-parametric significance test for the envisaged pairwise comparisons of the concepts' assessments by the independent sub-samples as well as for the comparison of the group of students and the group of adult sub-samples. Analogously to the aforementioned procedures in the second round, the test is carried out for each category and a significance level (p) of 0.05 is applied (cf. 3.2.2.3.2).

Comparison of the Concepts

With respect to the hypothesis regarding the views of the participants reflecting different accentuations of desirable science education (cf. hypothesis 1c), the third round yields data on the basis of which it can be investigated whether the concepts differ from each other in their assessments.

As normal distribution of the data of the third round is not likely, a non-parametric test has to be applied again. Moreover, for the purpose of pairwise comparison of the perceived priority and practice as well as the resulting priori-ty-practice differences of the concepts, significance tests for related samples have to be used. The Wilcoxon signed-rank test provides an appropriate proce-dure for investigating whether the concepts show significant differences in their perceived priorities, degrees of realization and the priority-practice differences, both regarding science education in general and different educational levels. As a non-parametric testing procedure for related samples, this test compares whether they differ, on the basis of median values, in their central tendencies. Statistically significant outcomes of this test indicate that differences between the data of the compared assessments do not result from random distribution (Bortz, 2005, pp. 153-154). In this test, a significance level (p) of 0.05 is assumed as well.

4 Empirical Part

4.1 Results of the First Round

In the following sections, I will describe the results of the first round of the Berlin Curricular Delphi Study in Science. In view of the both qualitative and quantitative nature of the data, the presentation of the results is divided into two parts. First, I will address the results of the qualitative analysis, including the inter-rater agreement, the classification system and a characterization of the responses. These outcomes are followed by the results of descriptive statistical analyses such as response behavior and category frequencies.

4.1.1 Description of the Sample and Response Rate

First Attempt

Between February 2011 and June 2011, I asked more than 1000 stakeholders via mail and e-mail to take part in the Berlin Curricular Delphi Study in Science and to fill in the questionnaire of the first round. The sample structure, the number of participants in each group, and the response rate after the first attempt are displayed in Table 2.

As shown, 140 out of 1030 experts initially followed the request to take part in the first round of the Berlin Curricular Delphi Study in Science. The largest number of returned questionnaires belongs to the group of scientists with N=59, followed by the group of teachers (N=41). Students (N=25) and science education researchers (N=15) have shown rather low willingness to participate. As expected (cf. 3.1.2), none of the contacted participants from the group of education policy took part in the first round of the Berlin Curricular Delphi Study in Science Education.

Table 2
Sample Structure and Response Rate in the First Round After the First Attempt

Sample Round 1		Number of Stake-holders contacted	Number of Responses		Response Rate
Students		50	25		50%
Teachers	Education Students	40	16		
	Trainee Teachers	20	4		
	Teachers	45	14	41	21%
	Teacher Educators	90	7		
Education Researchers		35	15		43%
Scientists		750	59		8%
Education Policy		10	0		0%
Total		1030	140		14%

Second Attempt

With respect to the targeted sub-sample composition (cf. 3.1.2) needed for the intended type of differentiated analyses (cf. 3.2.1.3) and anticipated dropout that is likely to occur in long-term multi-stage studies (cf. 2.3.3.1.8), the composition of the sub-samples was unsatisfying or – in case of the education researchers – even insufficient after the first attempt. Hence, a second attempt was necessary. For this purpose, further stakeholders potentially willing and suitable to take part in this study were contacted between August 2011 and March 2012 and asked to take part. In this way, more than 1100 stakeholders were contacted within the first round of this study. The final structure of the sample in the first round and the response rate after the second attempt are shown in Table 3. As can be seen, the number of participants has increased considerably after the second attempt. Reasons for this increase might be connected to a more thorough selection of further potential participants.

Finally, a total of 193 participants have participated in the first round of this study. The teachers (N=63) are the largest group (33% of the total sample), followed by the scientists (N=61), who represent 32% of all participants. The group of students increased to N=39 (20%) and the group of education researchers reached a size of N=30 (16%). As the Delphi method is based on a fixed group of participants throughout the different rounds (cf. 2.3.3.1.3), the group of education policy can due to lack of participation in the first round not be contacted for participation in the further procedure of this study. A detailed overview of the sample structure is given in Table 4.

Table 3
Sample Structure and Response Rate in the First Round After the Second Attempt

Sample Round 1		Number of Stake-holders contacted		Number of Responses		Response Rate
Students		70		39		56%
Teachers	Education Students	60		32		
	Trainee Teachers	25	235	5	63	27%
	Teachers	55		18		
	Teacher Educators	95		8		
Education Researchers		60		30		50%
Scientists[16]		750		61		8%
Education Policy[17]		10		0		0%
Total		1125		193		17%

---

16 With respect to the large sample size in the group of scientists after the first attempt, no new participants of this sub-sample were contacted in the second attempt. The two additional participants of this group included after the second attempt result from belated responses from the first attempt.
17 Concerning the group of education policy, all contacted stakeholders from the first attempt were contacted again in the second attempt.

Table 4
Detailed Structure of the Final Sample of the First Round

Group	Sub-Group und Subjects		Distribution[18]	Total Number	Percentage
Students	Basic Science Courses	Biology	7	39	20.0%
		Chemistry	6		
		Physics	6		
	Advanced Sciences Courses	Biology	2		
		Chemistry	14		
		Physics	4		
Science Teachers	Science Education Students at University	Biology	10	32	32.6%
		Chemistry	17		
		Physics	3		
		Science (elementary)	2		
	Trainee Science Teachers	Biology	1	5	
		Chemistry	3		
		Physics	1		
		Science (elementary)	0		
	Science Teachers	Biology	3	18	63
		Chemistry	4		
		Physics	5		
		Science (elementary)	0		
		not specified	6		
	Trainee Science Teacher Educators	Biology	2	8	
		Chemistry	4		
		Physics	2		
		Science (elementary)	0		
		not specified	0		
Science Education Researchers		Biology	8	30	15.5%
		Chemistry	13		
		Physics	6		
		Science (elementary)	1		
		not specified	2		
Scientists		Biology	8	61	32.6%
		Chemistry	9		
		Physics	26		
		Others	18		
Education Policy		Repr. of ed. policy	0	0	0.0%
		Formal ed. authorities	0		
Total				193	----

18 As multiple answers concerning these characteristics were possible, the numbers in this column
do not represent the summands of the total numbers of participants in the respective sample
groups.

4.1.2 Qualitative Analysis

I will describe the results of the qualitative analysis of the open-text material through the inter-rater agreement, which represents the degree of homogeneity in the coding of the participants' statements, and in terms of the final classification system. Moreover, I will provide a characterization of the different sub-samples' responses, which is enhanced by several examples of open-text responses.

4.1.2.1 Inter-Rater Agreement

In order to analyze the participants' responses, their statements were coded according to principles of content analysis (cf. 3.2.1.3.1). In order to investigate the degree of consensus in the coding of the participants' responses and address in this way the issue of coding objectivity, the inter-rater agreement between two raters was determined after coding 20 questionnaires on the basis of the following formula according to Bolte (2003b, p. 15), Häußler et al. (1980, p. 115) and Mayer (1992, p. 61):

$$q = \frac{2N_+}{2N_+ + N_-}$$

N_+: Number of cases in which positive category coding of both raters matches
N_-: Number of cases in which only one rater assigns a category

Table 5
Results of the Inter-rater Agreement After Coding 20 Questionnaires

I:	IIa:	IIb:	III:	IV:
Situations/ contexts/ motives	Concepts and topics	Fields and perspectives	Qualification	Methodical aspects
q_I=0.78	q_{IIa}=0.82	q_{IIb}=0.70	q_{III}=0.74	q_{IV}=0.76
		q_{total}=0.77		

As can be seen in Table 5, the section-based quotients of the inter-rater agreement range between 0.74 and 0.82. The overall inter-rater agreement is q=0.77. With regard to the question of benchmark values on the basis of which "good" inter-rater agreement can be assumed, different thresholds can be found in the literature (Greve & Wentura, 1997, p. 111). With reference to a classification of kappa values according to Landis and Koch (1977, p. 165), values between 0.61 and 0.80 are characterized as indicating substantial agreement, values ranging between 0.81 and 1.00 as showing almost perfect agreement. Fleiss (1981, p. 218) even characterizes values over 0.75 already as excellent and values between 0.40 and 0.75 as fair to good. Similarly, Bortz and Döring (2006, pp. 275-277) mention a threshold between 0.60 and 0.75 to indicate good inter-rater agreement. With respect to a minimum value, they argue that inter-rater agreement should not fall below a value of 0.4 in order to speak of sufficient reliability. Following these considerations, it can be said that values over 0.75 can be considered as good to excellent, and values between 0.40 and 0.75 are acceptable. According to these criteria, the extent to which the two coders allotted the open-text answers of the participants' responses to the same categories can be considered a satisfying inter-rater agreement. In this way, the established category system and the applicability of its usage for final labeling and coding of all statements as well as its suitability as an adequate representation of the participants' responses is confirmed

4.1.2.2 Classification System

An analysis of the open-text answers provided by the stakeholders in the first round of this study according to the procedure described in 3.2.1.3.1 yields a classification system of 88 categories. The classification system is divided into different content sections (Table 6). This partition follows a division by Bolte (2003b) and refers to the structure of the questionnaire in terms of situations, contexts, motives (I), field (II), and qualification (III). Part IV (methodical aspects) was established as an additional part. For part I (situations, contexts, motives) 18 categories were developed, relating to individual education (N=3), external motives (N=2), situations (N=2), individual (everyday life related) contexts (N=7), and scientific contexts (N=4). Part II (field) contains a total of 44 categories. The sub-parts IIa (basic concepts and topics) and IIb (scientific fields and disciplines) consist of 20 and 24 categories respectively. They relate to basic concepts of science (N=11), basic topics of science with reference to everyday life (N=9), perspectives of the sciences (N=16), and perspectives from which science as well as everyday life related facts can be considered (N=8). Part III (qualification) consists of 18 categories which refer to general attitudes (N=3), everyday life-related (scientific) competencies (N=11), and scientific and inquiry-related competencies (N=4). The additionally established part referring to methodical aspects (part VI) contains 8 categories.

As the category system in this study emerges from an analysis that aims at displaying participants' category references in terms of statement bundles (cf. 3.2.1.3.1), the established categories are not necessarily independent from each other but are interrelated and may also partially overlap.

Table 6
Overview of the Categories for the Analysis of the Stakeholders' Statements

I: Situations, Contexts, Motives N=18	II: Field		III: Qualification N=18	IV (Addition): Methodical Aspects N=8
	IIa: (Basic) Concepts and Topics N=20	IIb: Scientific Fields and Perspectives N=24		
- Education/ general pers. development	- Matter/ particle concept	- Botany	- Motivation / inter- est/curiosity	- Coopera- tive learning
- Emotional pers. devel.	- Structure/ function/ properties	- Zoology	- Critic. questioning	- Learning in mixed- aged classes
- Intellectual pers. devel.	- Chemical reactions	- Human biol.	- Acting reflected- ly/responsibly	
- Students' interests	- Energy	- Genetics / molecular biology	- Knowledge about sc. occupations	- Interdis- ciplinary learning
- Curriculum framework	- System	- Microbiology	- Content knowledge	- Inquiry- based science learning
	- Interaction	- Evolutionary biology	- Comprehension/ understanding	
- Me- dia/current issues	- Develop- ment/ growth	- Neurobiology	- Applying knowledge/ thinking abstractly	- Learning at sta- tions
- Out-of-school learning	- Models	- Ecology	- Judgement/opinion- forming/reflection	- Role play
	- Terminology	- General and Inorganic chemistry	- Working self- dependent- ly/structuredly/ pre- cisely	- Discus- sion / debate
- Na- ture/natural phenomena	- Sc.inquiry	- Organic chemistry	- Finding information	- Using new me- dia
- Everyday life	- Limits of sc. knowledge	- Analytical chemistry	- Reading compre- hension	
- Health	- Cycle of matter	- Biochemistry	- Communication skills	
- Technology	- Food/nutriti on	- Mechanics	- Soc.skills/teamwork	
- Society/pub. concerns	- Medicine	- Electrody- namics	- Sensibility/empathy	
- Global refer- ences	- Matter in everyd. life	- Thermody- namics	- Perception/ aware- ness/ observation	
- Occupa- tion/career	- Technical devices	- Atomic/ nuclear phys.	- Formulating scien- tific ques- tions/hypotheses	
	- Environ- ment	- Astronomy/ space system	- Experimenting	
- Science - biology	- Industrial processes	- Earth sc.	- Analyzing / drawing conclusions	
- Science - chemistry	- Safety and risks	- Mathematics		
- Science - physics	- Occupations /occ. fields	- Interdisc.		
- Science – interdiscipli- narity		- Current scientific res.		
		- Consequences of technologi- cal develop- ments		
		- History of the sciences		
		- Ethics/values		

4.1.2.3   Characterization and Examples of the Responses

Students

An analysis of the statements in the group of students shows that in many cases, the responses are generally short, referring to keywords or including concise formulations rather than elaborate and longer sentences. However, some students also provided comprehensive full-text answers. In general, the suggested contexts, motives, situations, as well as contents and qualifications in the group of students can be characterized by a strong affectedness by a variety of references to their everyday lives. In several cases, students mention contexts and topics that appear frequently in groups of young people, such as cosmetics, smoking, drug issues, and alcohol consumption.

> "The number of cases concerning alcohol abuse has increased. Therefore, it is important to understand its structure, species and negative consequences / effects to learn how to deal with alcohol appropriately"[19]. Student_111_4_fs1

Also, students mention the aim to better understand functions and processes of the human body and learn about related aspects such as diseases and medicine, nutrition, and healthy lifestyle.

> "Aspects which are important for a healthy life, such as nutrition, muscle functions of the human body, processes of the human body (biology)." Student_112_2_fs1

> "Understanding one's own body." Student_111_7_fs1

> "Science education should include current and everyday topics, e.g. the composition of medicine, food, etc." Student_122_11_fs1

However, the students' responses are characterized by several references to content knowledge as well:

> "The goal should be at least basic knowledge in the science subjects." Student_112_3_fs1

> "At the end of compulsory education, one should know roughly how electricity works (in daily life), how food preparation takes places, which processes occur in the human body, what air is composed of, what the meaning of acidic and basic is, and how nature works." Student_122_1_fs1

The appearance of these two approaches in the students' statements suggests that the students are not content with everyday life contexts only, nor do they see

---

19   All of the provided excerpts are translations by the author of this study.

science as a collection of detached facts, but wish to experience science classes that combine both everyday life and science-specific content knowledge.

In many responses, experimenting is explicitly stressed as a central aspect of science education and elements connected to the area of scientific inquiry are proposed:

"Much practical illustration through experiments." Student_ 112_1_fs1

"Skills related to carrying out and evaluating experiments independently should be developed." Student_112_1_fs1

"Scientific methods." Student_113_5_fs1

"The abilities to analyze, interpret and understand, to describe." Student_122_6_fs1

"Practice based science classes are more interesting for the students." Student_113_1_fs1

In addition, several responses of the students refer to incidents which appeared as important topics in the news during the time of the first round and that can be approached from a scientific perspective.

"It would be nice to address current issues, although this is not always easy in science. However, an example would be the nuclear accident in Japan to teach young people about nuclear power." Student_112_3_fs1

"Explaining natural catastrophes and making them more comprehensible." Student_112_4_fs3

„Media coverage, current topics (e.g. EHEC bacteria)." Student_113_2_fs1

Further examples of such references include accidents in (chemical) industry, oil spills, hurricanes, diseases, global warming, floods, and pollution.

In many cases, the students' responses are formulated on an interdisciplinary level:

"Everyone should know how one's body works and what chemical and biological processes take place in the human body." Student_122_1_fs1

"Environment (gravity, energy generation), human body (structure, function, chemical processes, physiological stress, consequences)." Student_122_4_fs1

„In general, science [...] should be connectable to other subjects." Student_112_2_fs1

These extracts indicate that students do not strictly formulate their answers according to individual subjects, but rather connect different areas of science.

All in all, the examples show that the students' responses feature a strong relation to their personal and everyday lives, as well as to topics in media coverage, content knowledge, and scientific inquiry.

Teachers

An investigation of the open-text answers provided by the teachers reveals that in general, their responses are characterized by a large variety of aspects and elaborate rationales. Their responses include a high number of references to contexts, situations, topics, perspectives, and qualifications that are connected to overarching public concerns.

> "Fossil fuels are becoming more expensive, CO_2 emissions are harmful to the environment and nuclear energy entails the risk of a meltdown. Yet, more and more energy is consumed. It is important to know the disadvantages and advantages of the various energy sources." Teacher_253_3_fs1

Also, the significance of relating science education to everyday life is emphasized. At the same time, however, it becomes apparent that everyday life is not seen as content itself, but as a framework in which content knowledge is acquired:

> "Science education should build on the everyday life experiences of the students. Scientific phenomena / knowledge have to be placed into existing structures / experiences and expand those." Teacher_262_3_fs1

Moreover, many statements are about the importance of including elements from the field of scientific inquiry in science lessons:

> "[…] Being able to formulate a clear research question […]. Being able to name parameters by means of which the research question could be investigated (hypotheses, planning experiments, evaluation of experiments). Being able to analyze a phenomenon and provide though appropriate measures verifiable proofs." Teacher_265_2_fs1

> "Science education provides optimal opportunities to train problem solving thinking skills. Therefore, it is important in science education to initially learn and practicing processes of solving a scientific problem […] within the example of documented experiments as well as to develop these processes up to working independently on theoretical questions in order to be able to apply in all areas of life strategies and approaches of problem solving." Teacher_252_2_fs4

With respect to relevant contexts and topics to enhance students' scientific literacy, rationales are given in particular from the field of media coverage and exemplified with current issues:

"Current issues, e.g.: A major environmental disaster as currently happened in Japan. The earthquake and the tsunami in relation to the disaster at the Fukushima nuclear power plant can be used as a starting point to discuss the problems of the peaceful use of nuclear power (from possible accidents to reprocessing and repository). In this frame, the importance and possibilities of generating and using energy can be addressed more generally." Teacher_251_1_fs

"Through referring to current incidents, e.g. environmental issues (global warming, ozone layer, acid rain, dying forests, water eutrophication, etc.), hazards from specific industries: CO pipeline [...], poisoning of water in neodymium production in China (for the production of wind turbines), the use of alternative energy sources, solar, renewable raw materials (rapeseed, grain, generation of bioethanol, E10 fuel, CO_2 balance), students should be encouraged to be interested in scientific issues." Teacher_253_4_fs1

In addition, special emphasis is placed on abilities related to critical questioning, judgment, opinion-forming, and reflection:

"Assessing tabloid headlines such as 'killer germ is killing all of us', 'Fukushima-killer dust comes to us', or 'toxic gas explosion at chemical plant." Teacher_262_2_fs3

"The threat posed by the term chemistry: Food full of chemistry, chemistry = unhealthy, chemistry as the opposite of 'natural' or 'organic'" Teacher_262_2_fs4

"Critical but unbiased attitude towards science." Teacher_250_3_fs1

In some cases, teachers also provide lists with suggested topics. In the context of climate change, these include, for example:

"Composition of the air, identification of gases, alkanes, fossil fuels, greenhouse effect, ozone hole." Teacher_262_2_fs1

Like the group of students, though in more elaborate examples, teachers make multiple references on an interdisciplinary level:

"Interest in physics diminishes rapidly with increasing age. Moreover, many students are often not able to explain simple physical phenomena. This goes hand in hand with the linking of the scientific disciplines. With respect to the students' ages [...], it would be possible, e.g. to address the topic car / motor scooter on an interdisciplinary level, that is, to work on this topic with approaches from all science subjects [...] As mentioned above, science education should be interdisciplinary. [...] This also promotes the ability to consider an issue from different viewpoints." Teacher_250_6_fs1

Moreover, the teachers' responses are characterized in particular by aspects of social responsibility, sustainability, and references to global developments. Also, some teachers express their concerns about current science related social issues, as the following excerpts show.

> "Another important issue refers to current societal problems which give rise to important scientific issues, currently issues such as Fukushima, E10, or continuing problems, such as lack of water or pesticides." Teacher_242_2_fs1

> "Long-term trends as a framework for the improvement of human life on earth are contexts to be selected for science education in terms of scientific literacy. Mega trends are: population boom, use of resources, climate change, globalization." Teacher_251_2_fs1

In these examples, the society's responsibility for the environment and for eco systems worldwide is emphasized. In such contexts, especially the discrepancy between the current exploitation of resources and a responsible use and conservation is mentioned.

Science Education Researchers

A consideration of the responses provided by the science education researchers reveals that their responses are even more than in the group of teachers characterized by a large variety of aspects and very sophisticated and elaborate statements. In the same way as in the group of teachers, "everyday life" is one of the key terms in the statements of the science education researchers and described as a central frame of science education.

> "References to the living environment of young people and taking up problems of everyday life. Content that might be relevant in their future careers and everyday life, e.g. health education, ecology, sustainable development, evolution, genetics and genetic engineering, energy and environmental issues, bionics, ethics in science, nature of science." Sc. Ed. Researcher_3995_6_fs1

> "It would be important to connect teaching more closely to the lives of the students and include scientific topics / phenomena of everyday life. There are many contexts from the news / media that can function as motivational aspects for that purpose. It is also important [...] that they become aware of how they can apply their knowledge in everyday life." Sc. Ed. Researcher_3995_8_fs1

Scientific inquiry as a central characteristic of science is emphasized by the science education researchers in several statements:

> "Young people should [...] acquire fundamental scientific skills, understand how the prototypical process of sc. inquiry takes place."Sc.Ed.Researcher_3992_3_fs5

> "[...] Scientific methods (investigating, observing, experimenting, etc.)." Sc. Ed. Researcher_3991_1_fs1

Science education researchers also set a strong focus on aspects that refer to social problems, public concerns, natural phenomena, environment, ecology, and current media coverage. Prominent issues within these areas include exploitation of resources, sustainability, global responsibility, and topics related to human biology.

"Social injustice in the globalized world and the destruction of valuable ecosystems require the integration of concepts of sustainable development in the classroom." Sc. Ed. Researcher_3991_2_fs3

"Due to the extinction of species, globalization, [...] on this planet, [...] students should learn about species and physiological processes as well as about what they consume [...]. Genetics and thus genetic engineering should not only be discussed in theory but should be critically assessed and discussed. In fact, education for sustainable development should be provided in school." Sc. Ed. Researcher_3995_1_fs1

"In the media, children and young people are regularly confronted with local or global environmental problems resulting from the use of human resources / certain economical ways. Often, scientific and technical knowledge is necessary to assess these issues and develop possible solutions. E.g.: Pollution of water, soil and air, risks of the use of nuclear energy, ozone hole, CO_2 / climate change." Sc. Ed. Researcher_3992_2_fs3

Moreover, science education researchers strongly underline everyday life-related (scientific) skills that are crucial within their content-related emphases. In particular, these include applying knowledge, and understanding as well as aspects from a social responsibility dimension such as judgement, opinion-forming, reflection, reflected and responsible acting.

"The competence to recognize problems and relationships, to question, to form assessments, and to develop opinions based on arguments, should be an outcome after 9-12 years of science education. The learner should be aware of his own role in the world, in the universe and in the social structure." Sc. Ed. Researcher_3995_1_fs1

"The students should be able to discuss scientific issues, master a basic technical terms, understand the context of complex situations such as climate change and be able to explain them, connect their knowledge, and assess scientific problems." Sc. Ed. Researcher_3995_8_fs1

"Using the acquired knowledge to understand ecological systems and developing an awareness of the sensitivity of the respective systems in terms of sustainable environmental education" Sc. Ed. Researcher_3992_4_fs4

"[...] Reflected and responsible acting towards oneself and others". Sc. Ed. Researcher_3992_2_fs4

When characterizing the science education researchers' statements, it also becomes apparent that within these emphases, the statements are mostly formulated on interdisciplinary levels:

> Areas of education and qualification: Biology / Chemistry / Geography: 1.) Scientists believe that global warming affects human health. 2.) Global warming influences the total production of food. 3.) Scientists expect more extreme variation of weather with devastating effects. Sc. Ed. Researcher_3995_1_fs3

As a very important goal of science education that aims at qualifying students as scientifically literate citizens, a general attitude of critical questioning is underlined. The following emphasis is paradigmatic for several responses from science education researchers.

> "A generally critical perspective on the basis of scientific inquiry in order to counter an all too gullible attitude towards information from all sectors of society ("type of person": (self-) critical reflected citizen)". Sc. Ed. Researcher_3993_5_fs1

In addition, the science education researchers' responses are characterized by their in-depth knowledge in science education research, which becomes apparent in the theoretical considerations and remarks they provide:

> "Scientific processes of understanding can be a key to independent thinking and thus be of overarching importance for a fruitful culture of learning [...]. Particularly important is a facilitation of meaningful, personality-effective learning processes (Seligman 2005, Wagenschein 1991, Rogers 1990, Zimmermann 2011)." Sc. Ed. Researcher_3994_1_fs1

> "Developmental tasks of adolescence should be addressed within the frame of science-related general education." Sc. Ed. Researcher_3992_3_fs1

> "As motives for science education, for example, the classic key issues following Klafki can be used. On the one hand, there are, of course, environmental issues. [...]." Sc. Ed. Researcher_3993_3_fs1

All in all, as shown by the included extracts, the science education researchers' responses can be described by a strong focus on social, global and ecological concerns. In particular, the society's responsibility for the environment and for humankind is stressed. Also, the responses are shaped by an interdisciplinary character of the suggested contents.

Scientists

In general, the responses from the scientists are enhanced with a large variety of examples and can be described as very pragmatic. Similarly to the group of teachers and science education researchers, most responses refer to everyday life frameworks, current issues in society, public life, and media and include elements of scientific inquiry. In most cases, the responses are formulated on interdisciplinary levels.

"Genetic engineering has mainly negative connotations in public. Therefore, it is important that through knowledge about genetic engineering, benefits are shown in addition to the risks so that genetic engineering can be assessed better by the individuals." Scientist_3991_2_fs2

"Another motive should be health. A conscious lifestyle, prevention and emergency aid are based on fundamental knowledge of the human physique and its conservation. Unnecessary hysteria about largely harmless germs but also underestimation of actual health risks should be reduced by an appropriate science education." Scientist_3991_3_fs1

"For an understanding of current issues within the frame of the major themes of scientific research, it is essential to know fundamentals and methods of scientific procedure such as verifying theories and models by measuring physical quantities as a form of objective observation." Scientist_4993_6_fs1

"Interdisciplinary thinking will become more important in the future. [...] Connections between math-physics-chemistry-biology should be more emphasized. [...] Lateral thinkers are more in demand in the future." Scientist_3995_10_fs1

Moreover, content knowledge as well as scientific and general competences are underlined. These include in particular analyzing, drawing conclusions, comprehension, understanding, acting in a reflected and responsible way, judgement, opinion-forming, reflection, and applying knowledge.

"In general, the student should be able to tackle problems analytically. For this purpose, a basic understanding of the measurement and identification of problems or phenomena must be developed in all subjects. For this purpose, appropriate techniques must be practiced that can be applied repeatedly. [...] Also, the ability to recognize the essence of a scientific problem should be trained." Scientist_3992_3_fs1

"Science education should serve as a basis that enables a person to conscious and independent decisions regarding lifestyle, nutrition, and health. This should [...] provide a basis for argumentation. The aim should be that the person can e.g. deliberately decide [...] and is able to form his or her own opinions independently of newspaper headlines and opinion-making media." Scientist_3991_3_fs1

"Students should be introduced to scientific methods and critical analysis and evaluation of experiments and observations in the sciences subjects." Scientist_3992_9_fs1

"A sense for what we strained environmental resources with [...] and what risk we take [...] in order to consume products". Scientist_3992_8_fs2

"In modern societies, a basic understanding of scientific thought should be developed that enables to understand the technical aspects that determine our lives, but also enables to see the limits of scientific approaches. Only in this way, competent and democratically legitimate social and political decisions are possible in the present and future." Scientist_4993_6_fs1

With a large number of physicists represented within the sub-sample of scientists, a particular strong focus is placed on scientific topics and viewpoints such as technology:

"More and more technical products have found their way into all life situations of humans. The basic scientific knowledge regarding technical products, however, declines more and more. This allows for fears to arise [...]." Scientist_4993_18_fs1

"Stronger links of the lessons to applications in everyday life (e.g. in mobile communication, information technology, etc.), understanding the function of devices, knowledge about methods of production and use of technology." Scientist_4995_15_fs1

"Structure, function and hazards of technical devices and systems (nuclear power plants, high-voltage lines)" Scientist_3995_18_fs1

All in all, the scientists' statements are enhanced by many examples and characterized by multiple references to everyday life, social issues and global concerns. Competences both in terms of general skills and scientific inquiry are strongly underlined. Further emphasis is placed on interdisciplinary approaches and technologically related content knowledge.

4.1.3 Quantitative Analysis

In order to gain a more differentiated overview of the empirical data, descriptive statistical analyses are applied. In the analyses, I focus on describing the response behavior and relative category frequencies, both regarding the total sample and the four sub-samples. This allows for more detailed information about the degree of differentiation in the statements of the participants.

4.1.3.1   Response Behavior

The response behavior of the stakeholders is described in terms of the following characteristic values (Table 7 and Table 8):

- number of form sheets filled in by the participants
- average number of form sheets filled in per person
- total number of statements[20] provided by the participants
- average number of statements provided per person
- total number of different statements provided by the participants
- average number of different statements provided per person

As shown in Table 7, a total of 400 form sheets were returned by the participants. However, most of the participants did not seize the possibility to express their views on the ten identical form sheets they were provided with, as more than two thirds of the participants (68.4%) used only one form sheet to formulate their statements). On average, two form sheets were used per person. Taking into account multiple category entries per person, more than 3100 statements, with an average of 16 statements per person, were provided by the sample. A consideration of the number of *different* categories references (not taking into account multiple category entries) yields a total of 2616 different category references in the responses of the total sample and an average of 14 different categories per person (Table 7). With respect to the different sub-samples, the group of science education researchers features the highest average number of different category references per person (M=21). Hence, this group is characterized by the highest degree of differentiation in terms of providing different category references. In contrast, the lowest number of references to different categories (M=9) can be found in the group of students.

---

20   The term 'statements' is in this context used in the sense of category references.

Table 7

Response Behavior of the Sample in the First Round After the Second Attempt

Sample Round 1		N (Form Sheets)	M (Form Sheets)	N (State-ments)	M (State-ments)	N (Different Statements	M (Different Statements)
Students		69	2	415	11	361	9
Teachers	Education Students						
	Trainee Teachers	142	2	1147	18	926	15
	Teachers						
	Teacher Educators						
Education Researchers		83	3	828	28	632	21
Scientists		106	3	769	13	697	11
Total		400	2	3159	16	2616	14

Note. N=Number, M=Mean

Table 8

Number of Form Sheets by the Participants in the First Round

Number of Form Sheets Used	Frequency	Percentage [%]	Cumulative Percentages [%]
1	132	68.4	68.4
2	16	8.3	76.7
3	11	5.7	82.4
4	10	5.2	87.6
5	13	6.7	94.3
6	3	1.6	95.9
7	1	0.5	96.4
9	1	0.5	96.9
10	5	2.6	99.5
13[21]	1	0.5	100
Total	193	100	100

---

21 As ten blank form sheets were provided, this participant used even more additional copies to provide statements.

The differences in the quantitative characterization of the response behavior (Figure 6) as well as in the degree of differentiation (Figure 7) of the four sub-samples are visualized by the following boxplots.

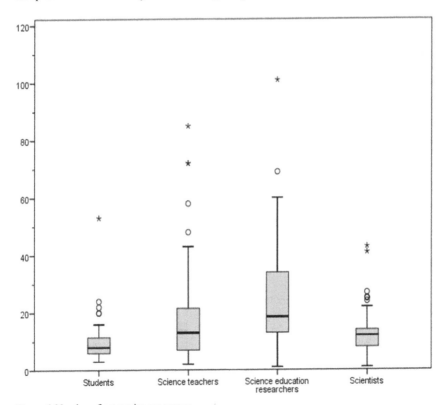

Figure 6. Number of categories per person

In both figures, it becomes apparent that the group of science education researchers features the most differentiated response profile, in terms of quantity as well as diversity. As indicated by the boxplots, 50% of the science education researchers refer in both cases to 18 or more (different) categories. In contrast, the number of aspects referred to by the students and scientists is considerably smaller. In case of making references to different aspects, the median values marked in the boxplots show that 50% of the students provided eight or less categories and 50% of the group of scientists provided 11 or less references. In the group of teachers, 50% of the participants provided 13 or more different category references. Hence, the group of teachers ranges in terms of differentiation in their responses below the group of science education researchers, but above the groups of students and scientists

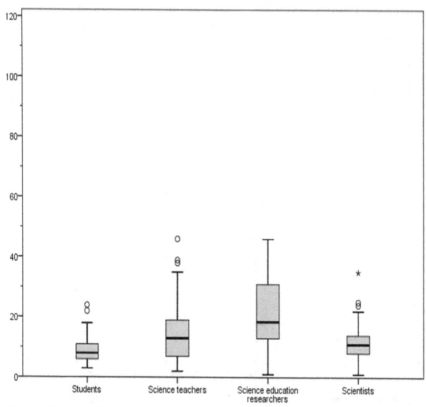

Figure 7. Number of different categories per person

4.1.3.2 Category Frequencies

In general, as indicated by the classification system, the participants' responses are characterized by a broad variety of aspects mentioned. In the following, the results of the qualitative analysis are complemented by an analysis of the category frequencies in the responses of the participants, representing a first approach to more detailed insights about the participants' views. When determining relative category frequencies, multiple entries by a person of the same category are not considered. In this way, empirical results are standardized. This is necessary because the participants used the given number of ten form sheets to very different extents (see 4.1.3.1). As the aim of this chapter is to illustrate the results of the first round in order to discover, in addition to general tendencies, as well distinctive emphases and most characteristic features in the responses of the participants, the analyses focus on categories mentioned particularly often or particularly rarely. For this purpose, aspects mentioned by more than 25% of the participants (f≥25%) as well as mentioned by less than 5% of the sample (f≤5%) are displayed.

4.1.3.2.1 Total Sample

In the following part, the frequencies of category references in the responses of the total sample are presented. The descriptions are structured according to the different parts of the classification system (cf. 4.1.2.2). Due to very incomplete and irregular entries regarding categories assigned to Part IV (methodical aspects), this section is not considered in the descriptions in order to avoid misrepresentation of the results. All of the discussed categories are also displayed in a bar chart (Figure 8) in descending order according to their frequency.

As Figure 8 shows, the categories "everyday life" and "media/current issues" are mentioned by a very high percentage of participants (59% and 50%) as important contexts to be considered in science education. Other categories of this section (part I) that are mentioned by more than 25% of the sample are "global references" (31%) and "society / public concerns" (31%). In contrast, only 4% of the participants mentioned aspects concerning the context of "curriculum framework".

In part IIa (concepts and topics), the aspect with the most references by the total sample is "scientific inquiry" (44%). Further categories mentioned often are "environment" (29%), and "medicine" (27%). "Occupational fields" and "development / growth" are only mentioned by 4% of the participants of the total sample. Concerning part IIb (disciplines and perspectives), the most frequently mentioned category is "human biology" (29%). "Earth sciences" (4%), "astronomy/space systems" (4%), "analytical chemistry" (3%) and "neurobiology" (2%) are mentioned only rarely.

With respect to qualifications (part III), a high number of categories are mentioned by more than 25% of all participants: "Acting reflectedly and responsibly" (39%), "content knowledge" (39%), "analyzing / drawing conclusions" (38%), "judgement / opinion-forming / reflection" (38%), "comprehension / understanding" (34%), "applying knowledge / thinking abstractly" (32%), and "critical questioning" (29%). "Knowledge about scientific occupations" (4%) and "reading comprehension" (3%) appear rarely.

All in all, a total of 15 categories are mentioned by more than 25% of the participants. 9 categories are mentioned by less than 5% of the participants.

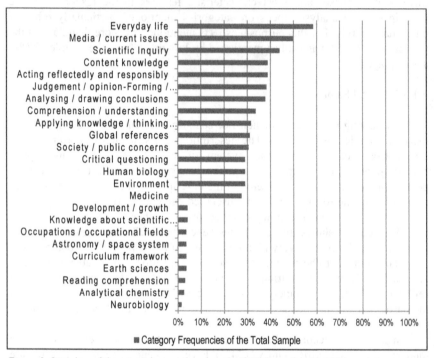

Figure 8. Overview of the categories mentioned rarely (f≤5%) or often (f≥20%) – relative percentages in the total sample

4.1.3.2.2 Sub-Samples

In the following part, a more differentiated look is taken at the category frequencies by focusing on the four sub-samples.

Students

As indicated in Figure 9, it can be seen that in the group of students, a total number of 8 categories was mentioned by more than 25% of the participating students. 13 categories are mentioned by less than 5% of the students and 9 categories are not mentioned at all. Hence, there are more categories mentioned to a low extent or not at all than categories referred to at a higher percentage than 25%. The most frequently mentioned categories in this group include "everyday life" (46%), "content knowledge" (44%), "scientific inquiry" (44%), and "media/current issues" (41%). Aspects related to "reading comprehension", "finding information", "astronomy / space system", "thermodynamics", "occupations/occupational fields", "structure / function / properties", "science – physics", "science - chemistry", and "curriculum framework" are not mentioned.

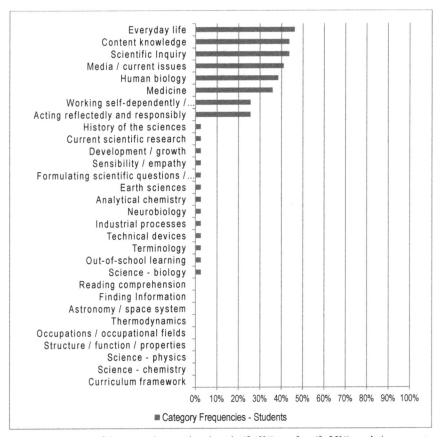

Figure 9. Overview of the categories mentioned rarely (f≤5%) or often (f≥25%) – relative percentages in the group of students

Teachers

In the group of teachers, 21 categories are mentioned often, while 16 categories are referred to by less than 5% of the teachers (Figure 10). Three categories are mentioned by more than half of the sub-sample. These include "everyday life" (65%), "scientific inquiry" (57%), and "media / current issues" (54%). Moreover, it can be seen that in this group, all categories are covered.

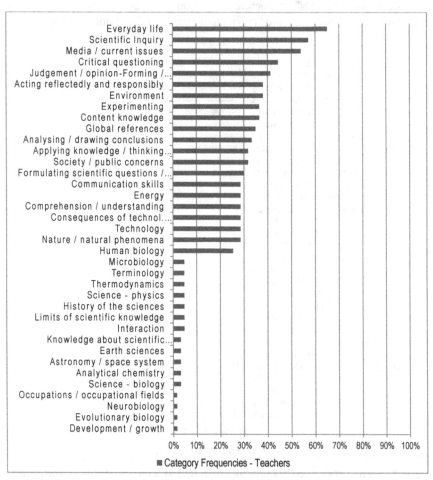

Figure 10. Overview of the categories mentioned rarely (f≤5%) or often (f≥25%) – relative percentages in the group of science teachers

Science Education Researchers

In the group of science education researchers (Figure 11), 33 categories are mentioned by more than 25% of the science education researchers.

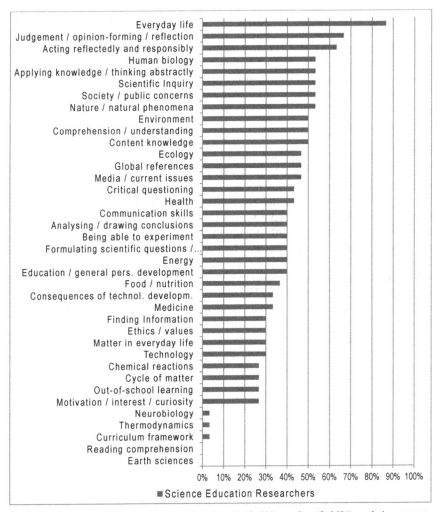

Figure 11. Overview of the categories mentioned rarely (f≤5%) or often (f≥25%) – relative percentages in the group of science education researchers

In contrast, only 3 categories are referred to by less than 5% of the participants in this group and two categories are no mentioned at all. 11 categories are mentioned by at least half of this group. These include "everyday life" (87%), "judgement / opinion-forming / reflection" (67%), "acting reflectedly and responsibly" (63%), "human biology" (53%), "applying knowledge / thinking abstractly" (53%), "scientific inquiry" (53%), "society / public concerns" (53%), "nature / natural phenomena" (53%), "environment" (50%), "comprehension / understanding" (50%), and "content knowledge" (50%). The rarely mentioned aspects (f ≤5%) refer to "neurobiology" (3%), "thermodynamics" (3%), and "curriculum framework" (3%). Aspects related to "reading comprehension" and "earth science" are not mentioned in this group.

Scientists

In the group of scientists (Figure 12), 13 categories are mentioned by more than 25% of the participants in this group. In contrast, 16 categories are mentioned rarely (f≤5%) and 4 categories not at all. Similar to the group of students, the responses in this sub-sample are characterized by more degrees of category references below 5% than category references above 25%. Two categories are referred to by more than half of this group (52%): "analyzing / drawing conclusions" and "media / current issues". Further most frequently mentioned aspects include "everyday life" (46%), "comprehension / understanding" (41%), "acting reflectedly and responsibly" (36%), "content knowledge" (33%), and "judgement / opinion-forming / reflection" (31%). Aspects not mentioned refer to "analytical chemistry", "general and inorganic chemistry", "neurobiology", and "curriculum framework". Interestingly, aspects related to "occupation" as a relevant context for science education as well as references related to "occupational fields" as an important topic of science education are only mentioned rarely (2%) by the scientists who work as employees or even employers in scientific occupations.

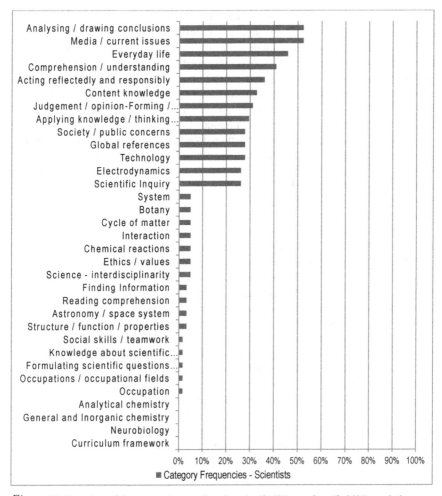

Figure 12. Overview of the categories mentioned rarely (f≤5%) or often (f≥25%) – relative percentages in the group of scientists

Comparison of the Sub-Samples

The category frequencies further illustrate the differences in the degree of differentiation between the four sub-samples. In general, it can be seen that the four groups reveal rather different frequency patterns. The students feature the highest number of rarely mentioned categories. At the same time, the number of frequently stated categories by the students is lower than in all the other groups. The group of teachers is the only group in which all categories are mentioned. In terms of frequently mentioned categories, they feature a higher number than the group of students and scientist, but a lower number than the group of science education researchers. The science education researchers include the highest number of categories with references over 25%. At the same time, they feature the lowest number of rarely or not mentioned categories.

When comparing frequently mentioned categories of the four sub-samples, it becomes apparent that 5 categories are shared by all the four groups among their often mentioned categories ($f \geq 25\%$): "media / current issues", "scientific inquiry", "content knowledge", "everyday life", and "acting reflectedly and responsibly". Such similarities cannot be found among the rarely mentioned categories ($f \leq 5\%$).

The chi-square test reveals that in the majority of cases, the groups do not differ significantly in their response behavior in terms of category references. Yet, for 34 out of 80 categories, significant differences between the four groups were found. This result indicates that in fact, for almost half of the categories, the four sub-samples express different emphases in the first round.

The "adult" sub-samples (teachers, science education researchers, and scientists), share even 12 frequently mentioned categories. They include (in addition to the above mentioned categories): "technology", "comprehension / understanding", "society / public concerns", "applying knowledge / thinking abstractly", "analysing / drawing conclusions", "global references", and "judgement / opinion-forming / reflection". Again, this observation cannot be replicated with respect to rarely mentioned categories. Comparing the group of students and adult sub-samples, the results of the chi-square test reveal significant differences for 15 out of 80 categories (19%). Hence, in the frequency of the majority of categories references, the students do not differ significantly from the adult sub-samples.

4.1.4 Discussion

As the first attempt of data collection featured a low response rate, especially in the group of science education researchers, students, and education policy, and thus did not provide sufficient distribution of participation in the different stakeholder groups, a second attempt had to be carried out to reach a solid sample size in view of the intended differentiated analyses as well as a balanced distribution of the participants among the sub-samples. All in all, after the second attempt, 193 experts took part in the first round of the Berlin Curricular Delphi Study in Science. As expected (cf. 3.1.2), none of the contacted participants in the group of education policy took part in the first round of the Berlin Curricular Delphi Study in Science Education. As all political parties and local administrative actors usually claim their demands in education issues, this result speaks volumes. Although the overall response rate was low, as can be anticipated in multi-round studies and long-term studies such as the Berlin Curricular Delphi Study in Science, the number and distribution of the participants of the four sub-samples is satisfying.

Following the procedure of statement analysis according to Bolte (2003b) and based on classifications from previous Delphi studies in science subjects as well as from research literature, a systematization of the participants' statements was reached through establishing a category system. With respect to demanding the processing of the statements in order to be as differentiated as necessary and as summarized as possible, the final classification system includes a total of 88 categories. Divided according to the structure of the questionnaire, the category sections refer to situations, contexts, motives (part I), basic concepts and topics (IIa), scientific fields and disciplines (IIb), and qualification (III). In order to distinguish methodical references from content based references, methodical aspects are established as an additional part. However, this part is due to very limited data not taken into account in the analyses and consequently, cannot be considered in further analyses either.

The results of the inter-rater agreement indicate satisfying agreement in the coding of two independent coders. This confirms the validity of the classification system for adequately displaying the participants' responses and thus, following the Delphi method, its applicability as a basis for both the quantitative analyses and the questionnaire of the following round. Therefore, as proposed in hypothesis 1a, it was possible to establish a suitable classification system for the purpose of representing the participants' opinions on the basis of the Delphi method.

In general, the referenced aspects are of rather interdisciplinary nature and touch a variety of social, global and environmental concerns, public issues, and everyday life matters as well as fundamental scientific topics. A large number of categories are comparable to aspects identified in previous curricular studies associated with science education as well as to systematizations in literature. For example, several of the categories in the context section (part I) are similar to contexts of PISA Science (OECD, 2007b). Further category compatibilities include both content and qualification sections in PISA Science (OECD, 2007b) and basic concepts in the German national science education standards (KMK, 2005a, 2005b, 2005c). As already shown by Bolte (2003b) with respect to desirable aspects of chemistry related science education, the identified aspects coincide to a great degree with criteria included in recommendations for modern science education and correspond to objectives of science education stated in position papers such as by the German Association for the Promotion of Math and Science Education (MNU, 2003) or the European Commission (Eurydice 2011, p. 68). Yet, as several additional aspects appear in the participants' responses, the identified aspects also enhance those recommendations and previous findings with further aspects and additional categories were established as well. These findings confirm hypothesis 1b: Many of the categories in the classification system established on the basis of the participants' responses correspond to objectives of (German) science education in literature and show considerable overlaps with categories from previous curricular Delphi studies in science subjects, but the established category system also includes additional aspects.

It is important to note that following the Delphi procedure, the categories of the classification system are not to be understood as building blocks on the basis of which scientific literacy can be composed in a modularized manner. Rather, the categories provide in the sense of statement bundles first insights into what aspects of science education are mentioned by a heterogeneous sample of stakeholders as meaningful in order to enhance scientific literacy. In this way, the classification system illustrates that a broad diversity of aspects related to contexts, motives, situations, concepts, topics, fields, perspectives, and qualifications are considered as important for science education in Germany. This makes clear that science education is not determined by single features, but shaped by a combination of a variety of comparably important aspects.

Further qualitative analysis of the open-text answers with focus on the sub-samples allows for additional insights into the type and character of the responses. In general, the science education researchers and teachers provided the most elaborate and sophisticated statements, which are likely to be driven by both their teaching experiences in science education and theoretical knowledge about science teaching. In terms of content, their responses can be characterized by an orientation towards sustainability and citizenry. Moreover, their responses are strongly shaped by aspects often referred to as "higher-order thinking skills", "quantitative reasoning", "estimation skills", "scientific attitude", "problem-solving skills", "decision-making skills", or summarized as "habits of mind" (AAAS,1993). The scientists' responses can be described as the most pragmatic approaches with respect to application of knowledge in the context of future professions and show strong references to scientific perspectives, which might be explained by their professional background, for example in areas of scientific research or in science-related fields in industry. The students' responses can be characterized as driven by a strong affectedness by aspects from their personal environment, their everyday lives and media coverage.

The category frequencies illustrate the emphases of the participants. Overall, the emphases are characterized by relevance to students' lives and present interests as well as to situations students may encounter later on at some point in their lives. An especially strong focus is set on aspects with references to personal life, media coverage, and more general overarching skills as well as scientific content. In particular, these include "everyday life", "media / current issues", "scientific inquiry", "content knowledge", "acting reflectedly and responsibly", "judgement / opinion-forming", and "comprehension / understanding". In many cases, these emphases correspond to science curricula and recommended priorities in education literature in Germany and beyond (AAAS, 1990, 1993, 1997, 2001; Council of Ministers of Education, Canada, 1997; EC, 2011; GDCh, 2005; KMK, 2005a, 2005b, 2005c; MNU, 2003; NRC, 1996, 2000). Several of the identified tendencies are also comparable to aspects found to be considered important in science education in earlier studies (Bennett, Lubben, & Hogarth, 2007; Bolte, 2003a; EC, 2011; Häußler et al., 1980; Osborne & Collins, 2000).

Categories mentioned rarely mostly relate to aspects in the field of specialized knowledge, scientific sub-disciplines, industry, and scientific occupations. As suggested by Bolte (2003b), low considerations of certain categories might indicate that the corresponding aspects playing an only minor role in the classroom so that without exemplifying references, representatives of sub-samples such as students or scientists were not aware of these aspects and thus were not able to take these categories into account in their open-text responses of the first round.

For the quantitative analyses, characteristic descriptive-statistical values and frequencies of category references provide more detailed insights into the response behavior of the participants. In general, the group of science education researchers features the most differentiated response profile with an average of 21 different category references per person, both in terms of quantity and diversity, followed by the teachers, the scientists, and finally students, who, in contrast, provided 9 different category references on average. The category frequencies of the four sub-samples show that several identical categories can be found among the most frequently mentioned categories of all sub-samples. In fact, the categories "media / current issues", "scientific inquiry", "content knowledge", "everyday life", and "acting reflectedly and responsibly" are shared by all four groups among their often mentioned categories ($f \geq 25\%$). Hence, on the basis of the most frequently mentioned categories, the sub-samples consider several identical aspects as desirable for science education and hypothesis 5a can be confirmed.

While the sub-samples broadly agree in their main tendencies, a more detailed consideration of the category frequencies suggests that the stakeholder groups deviate from each other in the relative frequency of referring to the individual categories and thus differ in their specific focuses. As significant differences appear between the four groups in 34 out of 80 categories (43%), the assumption of differences between the sub-samples in their specific views (hypothesis 5b) can only partly be confirmed for the first round. Concerning the comparison between the group of students and the groups of adult sub-samples, certain similarities seem to be shared by the adult groups as opposed to the students. However, the hypothesis of such generational conflict (hypothesis 5c) cannot be confirmed by the chi-square-test, as significant differences appear only for 15 out of 80 categories (19%). Hence, in the first round, no clear gap between the expectations of the young generation and the adult groups regarding aspects of desirable science education can be found. Possible reasons for the identified differences might be related to these groups' different ways of approaching the question of desirable science education. While the adults, especially the science education researchers, might argue from a more theory based perspective due to their professional backgrounds, the students might argue from a more school based point of view based on their own educational experiences. In this way, differences between students and the adult sub-samples might again also be related to the assumption that certain categories play an only minor role in the classroom so that the students were not aware of these aspects and consequently were not able to take these categories into account in their open-text responses of the first round.

All the more interesting is thus the question whether the categories mentioned only rarely are generally considered less relevant or are not taken into account by the participants in the first round for other reasons. Also, the question arises if the frequently mentioned aspects are actually realized to a correspondingly high extent in educational practice. These questions are investigated in the second round of the Berlin Curricular Delphi Study in Science.

4.2 Results of the Second Round

In the following parts, I will address the results of the second round of the Berlin Curricular Delphi Study in Science. First, I will describe the sample and the response rate. According to the two-part structure of the second round questionnaire (cf. 3.2.2.2), the presentation of the results is divided into two parts. The first part (4.2.2, 4.2.3, 4.2.4) refers to results from statistical analyses based on the assessments on the 6-tier rating scale and is subdivided according to priority, practice, and priority-practice differences. The analyses are carried out on the data basis of the total sample as well as the four sub-samples (students, teachers, educations researchers and scientists). In the second part (4.2.6), I will describe the results of a hierarchical cluster analysis of category combinations which the participants provided in the second task.

4.2.1   Sample and Response Rate

The distribution and return of the questionnaires of the second round, again in paper-and-pencil as well as electronic form, took place between October 2011 and May 2012. As the Delphi method is based on a fixed group of participants throughout the rounds, only the 193 participating stakeholders from the first round were contacted for the second round. Table 9 shows the sample structure and participation rate with regard to drop-out between the first and second round. It can be seen that out of the 193 participants from the first round, a total of 154 participants (80%) took part in the second round.

Table 9
Sample Structure and Response Rate in the Second Round

Sub-Sample		Number of Responses				Response Rate
		Round 1		Round 2		
Students		39		34		87%
Teachers	Education Students	32		29		91%	
	Trainee Teachers	5	63	4	50	80%	79%
	Teachers	18		16		89%	
	Teacher Educators	8		1		13%	
Education Researchers		30		29		97%
Scientists		61		41		67%
Total		193		154		80%

As shown in a detailed overview of the sample structure in the second round of this study (Table 10), the group of students makes up 22% of the total sample with 34 participants. The group of teachers consists of 50 participants (32%) and thus constitutes the largest sub-sample. The group of science education research-ers represents the smallest part of the sample, consisting of 29 participants (19%). The scientists include 41 participants and are with 27% of all participants the second largest part of the sample.

Table 10
Detailed Structure of the Sample in the Second Round

Sub-Sample	Sub-group	Subject	Distribution[22]	Total Number	Percentage
Students	Basic Science Courses	Biology	16		
		Chemistry	21		
		Physics	17	34	22.1%
	Advanced Science Courses	Biology	5		
		Chemistry	10		
		Physics	5		
Science Teachers	Science Education Students at University	Biology	12	29	
		Chemistry	26		
		Physics	1		
		Science (elem. level)	2		
	Trainee Science Teachers	Biology	2	4	
		Chemistry	3		
		Physics	0		
		Science (elem. level)	0		50	32.5%
	Science Teachers	Biology	5	16	
		Chemistry	9		
		Physics	3		
		Science (elem. level)	3		
	Trainee Science Teacher Educators	Biology	0	1	
		Chemistry	1		
		Physics	0		
		Science (elem. level)	0		
Science Education Researchers		Biology	2	29	18.8%
		Chemistry	14		
		Physics	8		
		Science (elem. level)	0		
		not specified	5		
Scientists		Biology	8	41	26.6%
		Chemistry	13		
		Physics	12		
		Others	11		
Total				154	---

---

22 As multiple answers per participant with regard to these characteristics were possible, the numbers in this column do not represent summands of the total numbers of participants in the corresponding sample groups.

4.2.2 Priority Assessments

4.2.2.1   Total Sample

The mean values of the priority assessments provide first insights into aspects that are by the total sample considered most and least important for scientific literacy based science education. In the following analyses, the different sections of the category system are merged together for a more comprehensive consideration of the result. Due to irregular and incomplete data in part IV (methodical aspects), this part is not included in the descriptions. The results are displayed in Figure *13* in descending priority. With the applied 6-tier rating scale of the questionnaire ranging from 1 ("very low priority") to 6 ("very high priority), the theoretical mean value of 3.5 can be interpreted as fairly important. It can be seen that except "history of the sciences" (M=3.4) and "astronomy / space system" (M=3.3), all categories are on average assessed by the participants above the theoretical mean of 3.5. Thus, they are considered as at least fairly important, covering the assessments "rather high priority", "high priority" or "very high priority". The highest mean value is assigned to "comprehension / understanding" (M=5.3). Further categories considered as highly important (M≥5.0) include "analysing / drawing conclusions" (M=5.2), "applying knowledge / creative and abstract thinking" (M=5.1), "judgement / opinion-forming / reflection" (M=5.1), "critical questioning" (M=5.1), "nature / natural phenomena" (5.1), "acting reflectedly and responsibly" (M=5.1), "working self-dependently / structuredly / precisely" (M=5.0), "motivation and interest" (M=5.0), and "perception / awareness / observation" (M=5.0). The lowest priorities range between rather high and rather low priority and include "knowledge about science-related occupations" (M=3.9), "thermodynamics" (M=3.9), "microbiology" (M=3.9), "curriculum framework" (M=3.9), "earth sciences" (M=3.8), "botany" (M=3.7), "analytical chemistry" (M=3.7), "zoology" (M=3.7), "emotional personality development" (M=3.6), "industrial processes" (M=3.6), "history of the sciences" (M=3.4), and "astronomy / space system" (M=3.3).

    In addition to the analysis of the mean values, further insights into the participants' assessments can also be gained through a consideration of the standard deviations of the different categories. The standard deviation represents the degree of agreement between participants and thus displays whether participants share a consensus about the priority of a given aspect. The highest standard deviation and thus lowest degree of consensus among the participants appears for "ethics / values" (SD=1.3) and "emotional personality development" (SD=1.3). The lowest standard deviation and therefore the highest degree of agreement appears for "comprehension / understanding" (SD=0.8), which is at the same time the category with the highest priority mean value (M=5.3). This means that the participants largely agree on assessing this category as highly important for scientific literacy based science education. When comparing the standard deviations of the ten categories with the highest priorities against the mean value of all standard deviations (M_{SD}=1.0), it can be seen that 9 out of these 10 categories range below this value. This indicates that the agreement among the participants about most highly prioritized aspects is comparatively strong. All in all, with standard deviations between 0.8 and 1.3, and 42 out of 80 categories (53%) not surpassing a standard deviation of 1.0, it can be said that the different stakeholders mostly consider the same aspects as most and least important.

In order to address possible changes in the participants' emphases or reconsiderations of views over the rounds, as facilitated by the Delphi method, and to investigate whether the frequencies of the first round are reflected in the priority assessments, the findings of the frequency analyses in the first round are related with the assessments of the second round. For this purpose, the most frequently mentioned categories in the first round (f≥25%) are compared to those categories that range above the mean value of all priority assessments (M=4.4). In the same way, those categories mentioned only rarely in the first round (f≤5%) are compared to those categories with priority values below 4.4. The comparison shows that, except for "global references", all categories that were mentioned by more than 25% of the participants in the first round are assessed with mean values above 4.4. Hence, the most frequently mentioned categories from the first round are endorsed in terms of their priority in the second round. Yet, it becomes obvious that several additional categories, which were not mentioned by a high percentage of participants (f≥25%) in the first round, are in the second round in fact considered as important for science education, for example "knowledge about scientific occupations", "development / growth" or "earth science" . A comparison between categories mentioned only rarely in the first round (f≤5%) and categories with priority values below 4.4 in the second round reveals that except for "reading comprehension", all of the rarely mentioned categories in the first round fall below an average of 4.4 in the priority assessments of the second round and are thus not considered as the most important, yet still as at least fairly important, aspects of science education.

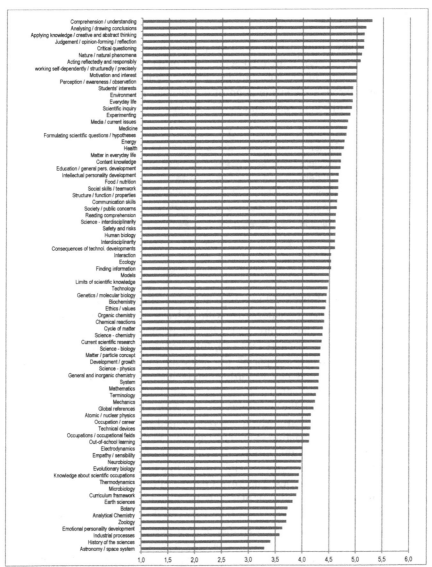

Figure 13. Priority assessments by the total sample[23]

23 This figure can also be accessed via www.springer.com and "Theresa Schulte" within the OnlinePLUS programme.

4.2.2.2   Sub-Samples

More differentiated information about the priority assessments in addition to the analysis of the priorities in the total sample are provided by a comparison of the priority assessments in the sub-samples. First insights into such a comparison can be gained through a consideration of the average priority assessments of all categories in the different sub-samples (Figure 14).

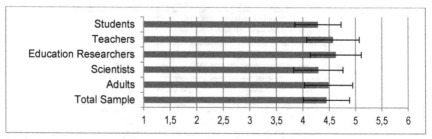

Figure 14. Overall mean values of the categories' priority assessments in the sub-samples

It can be seen that all mean values range above 4, which shows that on average, all sub-samples assess the categories with at least "rather high priority". The highest priority assessments are given by the group of science education researchers (M=4.6) and teachers (M=4.6), followed by the scientists (M=4.3) and students (M=4.3). When comparing the students and the group of adult sub-samples (M=4.5), a tendency towards slightly higher overall priorities can be identified in the assessments of the group of adult sub-samples. The degree of differentiation between the priorities of the individual categories is shown through standard deviations, which are represented by error bars in Figure 14. The highest standard deviation (SD=0.50) appears in the group of teachers, indicating that participants in this group differentiate to a higher extent between their assessments than the other groups with respect to assigning high and low priorities to the different categories. The lowest degree of differentiation in the priority assessments among the sub-samples can be found in the group of students (SD=0.44). Yet, with a difference of 0.06 between the highest and lowest standard deviation, the groups are very similar in their degree of differentiation.

More detailed comparisons of the priority assessments of the different sub-samples can be made on the basis of the results displayed in Table 11, Table 12, Table 13, and Table 14. The tables represent the different parts of category system (contexts motives and situations, basic concepts and topics, fields and perspectives, and qualifications).

Although in general, the different sub-samples assess the categories with similar priorities and mostly consider the same aspects as most and least important, the results of the significance test (Mann-Whitney-U-Test) indicate that the sub-samples differ in a number of pair comparisons from each other. All in all, a number of 117 statistically significant differences out of 480 possible pair comparisons of the four sub-samples can be identified. This represents 24% of all pair comparisons. The highest number of significant differences (N=27) appears between the students and education researchers and between the teachers and scientists (34% of all cases of these pair comparisons). The comparison between the students and teachers reveals 25 significant differences (31% of all cases in this group comparison), followed by the education researchers and scientists featuring 20 significant differences (25% of all cases in this group comparison). In the comparison between the students' and scientists' responses, 11 significant differences appear (13% of all cases in this group comparison). With 7 significant differences (8% of the cases in this pair comparison), the lowest number of significant differences appears between the responses of teachers and education researchers. A comparison of the students and the group of adult sub-samples reveals 25 significant differences (31% of the cases in this group comparison). In view of all pair comparison among the sub-samples, most significant differences appear in the part "qualification" (N=34).

As expectable in Delphi studies, the sample distribution in the second round is not equally balanced with respect to the four sub-samples. Thus, the results of the total sample are unproportionally composed by the sub-samples' assessments. In order to address this issue, the results are normalized to obtain values that are in equal parts composed of the mean values of each of the four sub-samples. A comparison of the normalized values with the original mean values of the total sample shows that except for two categories ("thermodynamics", $\Delta T=0.21$, and "analytical chemistry", $\Delta T=0.20$), all deviations range at absolute values of $\Delta T=0.04$ or below. In 65 out of 80 categories (81%), the differences between the original and normalized value range below $\Delta T=0.03$; 46 out of 80 deviations (58%) even fall below absolute values of $\Delta T=0.02$ (see appendix).

Table 11

Significant Differences (Mann-Whitney-U-Test) and Mean Values of the Priority Assessments Concerning Situations, Contexts and Motives

Category	Significant Differences							Mean Values				
	S/T	S/E	S/Sc	T/E	T/Sc	E/Sc	S/A	S	T	E	Sc	Total
Education / general pers. Development		*						4.4	4.7	5.1	4.7	4.7
Emotional personality development								3.3	3.8	3.9	3.6	3.6
Intellectual personality development								4.4	4.8	4.8	4.6	4.7
Students' interests		*			*	*		4.9	5.0	5.3	4.6	4.9
Curriculum framework								3.9	4.0	3.9	3.8	3.9
Out-of-school learning	*						*	3.6	4.4	4.1	4.2	4.1
Media / current issues								4.8	5.0	4.8	4.7	4.8
Nature / natural phen	*	*	*					4.6	5.3	5.3	5.1	5.1
Medicine							*	4.8	4.9	4.9	4.8	4.8
Technology	*		*				*	4.0	4.6	4.6	4.6	4.5
Everyday life	*	*			*	*	*	4.4	5.2	5.4	4.6	4.9
Society /public concerns					*			4.5	4.9	4.9	4.3	4.6
Global references					*			4.3	4.3	4.3	3.9	4.2
Occupation / career								4.2	4.2	4.2	4.1	4.2
Science - biology	*						*	4.6	4.1	4.3	4.4	4.3
Science - chemistry								4.5	4.2	4.4	4.4	4.4
Science - physics		*	*					4.0	4.3	4.3	4.6	4.3
Science – interdisc.	*						*	4.2	4.8	4.7	4.7	4.6
Number of stat. sig. differences	6	5	3	0	4	2	6					

$$\frac{6 \quad 5 \quad 3 \quad 0 \quad 4 \quad 2}{20} \quad 6$$

Note. S=students, T=teachers, E=education researchers, Sc=scientists, A=adults (teachers, education researchers, and scientists), *=significant ($p<0.05$)

Table 12
Significant Differences (Mann-Whitney-U-Test) and Mean Values of the Priority Assessments Concerning Basic Concepts and Topics

Category	Significant Differences							Mean Values				
	S/T	S/E	S/Sc	T/E	T/Sc	E/Sc	S/A	S	T	E	Sc	Total
Matter / particle concept				*	*			4.2	4.5	4.6	4.0	4.3
Structure / function / properties	*			*				4.4	4.9	4.8	4.4	4.7
Chem. Reactions	*			*				4.3	4.7	4.5	4.1	4.4
Energy								4.5	4.9	4.9	4.8	4.8
Development / growth								4.2	4.6	4.3	4.1	4.3
Models	*	*		*	*	*	*	4.1	4.8	4.8	4.2	4.5
Terminology			*	*	*	*		4.5	4.8	4.2	3.5	4.3
System								4.1	4.4	4.4	4.2	4.3
Interaction	*	*	*				*	3.9	4.8	4.7	4.6	4.5
Scientific inquiry		*			*			4.6	5.1	5.3	4.7	4.9
Limits of sc knowledge	*	*			*	*		4.2	4.7	5.0	4.2	4.5
Cycle of matter	*	*	*				*	3.9	4.6	4.6	4.3	4.4
Food / nutrition	*							4.3	4.9	4.8	4.5	4.7
Health								4.8	4.7	4.8	4.8	4.8
Matter in everyday life	*	*		*			*	4.3	5.0	5.0	4.5	4.7
Technical devices				*	*			4.1	3.8	4.4	4.4	4.2
Environment								4.9	5.0	5.0	4.8	4.9
Industrial proc.								3.6	3.5	3.5	3.7	3.6
Safety and risks				*				4.7	4.8	4.4	4.5	4.6
Occupations / occup. Fields								4.2	4.2	4.2	4.0	4.1
Number of stat. sig. differences	8	6	3	3	7	5	5					
			32									

Note. S=students, T=teachers, E=education researchers, Sc=scientists, A=adults (teachers, education researchers, and scientists), *=significant (p<0.05)

Table 13
Significant Differences (Mann-Whitney-U-Test) and Mean Values of the Priority Assessments Concerning Fields and Perspectives

Category	Significant Differences							Mean Values				
	S/T	S/E	S/Sc	T/E	T/Sc	E/Sc	S/A	S	T	E	Sc	Total
Botany		*		*		*		3.4	3.8	4.4	3.5	3.7
Zoology		*		*	*	*		3.3	3.9	4.4	3.3	3.7
Human biology			*		*	*		4.7	4.7	5.0	4.2	4.6
Neurobiology			*		*	*		4.4	4.1	4.1	3.5	4.0
Genetics / molecular biology								4.7	4.5	4.4	4.3	4.5
Microbiology	*							3.9	3.9	4.1	3.9	3.9
Evolutionary biology								4.1	4.0	4.2	3.7	4.0
Ecology	*	*			*	*	*	4.0	4.8	5.0	4.3	4.5
General and inorganic ch.					*			4.3	4.6	4.3	4.0	4.3
Organic ch..			*	*		*		4.6	4.7	4.2	4.0	4.4
Biochemistry					*			4.4	4.6	4.5	4.2	4.4
Analytical Chemistry			*				*	4.0	3.6	3.8	3.6	3.7
Thermodyn.								3.8	4.1	4.0	3.9	3.9
Electrodyn.								3.8	4.0	4.1	4.1	4.0
Mechanics								4.0	4.1	4.4	4.4	4.2
Atomic / nuclear physics								4.4	4.1	4.4	3.9	4.2
Astronomy / space system								3.1	3.1	3.8	3.3	3.3
Earth sciences								3.7	3.7	4.2	3.7	3.8
Mathematics					*			4.3	4.3	4.0	4.5	4.3
Interdisc.	*	*					*	4.0	4.8	4.9	4.5	4.6
Current scientific research			*		*			4.8	4.6	4.2	4.0	4.4
Consequences of technol.								4.6	4.6	5.0	4.3	4.6
History of the sciences								3.5	3.5	3.6	3.2	3.4
Ethics / values	*	*					*	3.9	4.7	4.7	4.4	4.4
Number of stat. sig. differences	4	5	5	3	8	6	4					
			31									

Note. S=students, T=teachers, E=education researchers, Sc=scientists, A=adults (teachers, education researchers, and scientists), *=significant (p<0.05)

Table 14
Significant Differences (Mann-Whitney-U-Test) and Mean Values of the Priority Assessments Concerning Qualifications and Attitudes

Category	Significant Differences							Mean Values				
	S/T	S/E	S/Sc	T/E	T/Sc	E/Sc	S/A	S	T	E	Sc	Total
Motivation and interest							*	5.0	5.1	5.2	4.8	5.0
Crit.questioning	*	*					*	4.8	5.3	5.3	5.1	5.1
Acting reflectedly/ respons.	*	*			*	*	*	4.7	5.3	5.4	4.8	5.1
Knowledge about science-related occup.								3.8	4.0	4.1	3.8	3.9
Content knowl.					*	*		4.8	4.9	4.9	4.4	4.7
Comprehension / understanding		*		*				5.1	5.3	5.6	5.2	5.3
Applying knowledge / creative/ abstract thinking	*	*			*		*	4.8	5.4	5.4	5.0	5.1
Judgement / opinion-forming / refl.		*					*	4.8	5.2	5.3	5.1	5.1
working self-dependently / structuredly / precisely								4.9	5.1	5.2	4.8	5.0
Finding inform.	*	*			.*		*	4.0	4.8	4.9	4.3	4.5
Reading comprehension	*	*			*		*	4.1	5.0	4.8	4.5	4.6
Comm. Skills	*	*					*	4.3	4.8	4.9	4.5	4.6
Social skills / teamwork					*	*		4.6	5.0	4.8	4.2	4.7
Empathy / sens.		*				*		3.6	4.2	4.4	3.7	4.0
Perce./ awareness / observ.	*	*			*		*	4.7	5.2	5.3	4.8	5.0
Formulating sc. quest. / hypoth.	*	*				*	*	4.5	4.9	5.2	4.6	4.8
Experimenting					*	*		4.8	5.1	5.0	4.5	4.9
Analysing / drawing concl.								4.9	5.2	5.4	5.1	5.2
Number of stat. sig. differences	8	11	0	1	8	6	10					
			34									

Note. S=students, T=teachers, E=education researchers, Sc=scientists, A=adults (teachers, education researchers, and scientists), *=significant ($p<0.05$)

4.2.3   Practice Assessments

4.2.3.1   Total Sample

The mean values of the practice assessments by the total sample reveal first insights into aspects that are considered as most and least present in the science classroom. As part IV (methodical aspects) features only irregular and incomplete data, this section is not included in the descriptions. The results of the total sample's practice assessments are displayed in Figure 15 in descending realization. On the basis of the 6-tier rating scale of the questionnaire ranging from 1 ("to a very low extent") to 6 ("to a very high extent), the theoretical mean value of 3.5 can be interpreted as fairly realized in science education. The results demonstrate that, in contrast to the priority assessments, only 22 out of 80 categories are on average assessed by the participants above the theoretical mean of 3.5. This shows that only few categories are considered as at least fairly present in science education. Instead, most aspects are considered as realized to a rather low or low extent and thus seen as being not very present in the science classroom. As opposed to the priorities, no category ranges above a practice mean value of 5. This means that no category is considered as being present to a high or very high extent. The aspect perceived as most present is "curriculum framework" (M=4.8). Further categories assessed as present in science education "to a rather high extent" (M≥4) include "content knowledge" (M=4.3), "chemical reactions" (M=4.2), "general and inorganic chemistry" (M=4.1), "terminology" (M=4.0), and "science – biology" (M=4.0). Comparing the ten highest priority values to the ten highest practice assessments, it can be seen that none of the priority top ten can be found in the practice top ten. Lowest degrees of realization include "occupations / occupational fields" (2.6), "limits of scientific knowledge" (M=2.6), "consequences of technological developments" (M=2.6), "out-of-school learning" (M=2.5), "ethics / values" (M=2.4), "knowledge about science-related occupations" (M=2.4), "emotional personality development" (M=2.3), "current scientific research" (M=2.3), and "astronomy / space system" (M=2.3). When comparing the ten lowest priority values to the ten lowest practice assessments, no matches are found. These comparisons yield first hints towards a discrepancy between wish and reality.

When considering the standard deviation as an indication for the degree of agreement among the participants in their assessments, it can be seen whether the participants share a consensus in their assessments of the realization of aspects in practice. The highest standard deviation and therefore lowest degree of consensus among the participants appears for "models" (SD=1.3) and "mathematics"

(SD=1.3). The lowest standard deviations and thus highest degree of agreement can be found for "media / current issues" (SD=0.9), "emotional personality development" (SD=0.9), "students' interests (SD=0.9), and "motivation and interest" (SD=0.9). When comparing the standard deviations of the ten categories with the highest practice assessments with the mean value of all standard deviations (M=1.1), it can be found that all of these categories but one ("environment", SD=0.9) are equal to or above this value. This shows only moderate agreement among the participants with respect to what aspects they consider as most realized in current science education. As the standard deviations of the ten lowest practice categories except for two aspects ("emotional personality development", SD=0.9 and "limits of scientific knowledge", SD=1.0) conform to the average standard deviations, the consensus among the participants regarding the least realized aspects is neither particularly strong, nor particularly weak. With a mean of $M_{SD}=1.1$ over all standard deviations, a slightly larger degree of disagreement can be found about the presence of these aspects in science education than regarding the priority of these aspects ($M_{SD}=1.0$). As the standard deviations of the practice assessments range between 0.9 and 1.3, and more than 38% of the categories (N=30) do not surpass a value of 1.0, the participants' assessments of the realization of aspects in science education broadly correspond to each other and in general, the different stakeholders mostly consider the same aspects as most and least realized in the science classroom.

In order to gain insights into how the category frequencies of the first round relate to the practice assessments, characteristic category frequencies of the first round are contrasted with the practice assessments of the second round. For this purpose, the most frequently mentioned categories in the first round ($f\geq25\%$) are compared to those categories that range above the mean value of all practice assessments (M=3.3). In the same way, categories mentioned only rarely in the first round ($f\leq5\%$) are compared to categories with practice values below 3.3. The comparison reveals that only three categories ("content knowledge", M=4.3; "environment", M=3.9; and "human biology", M=3.9) of those mentioned by more than 25% of the participants in the first round are in the second round assessed with mean values above 3.3. This indicates that only few aspects of the most frequently mentioned aspects are considered at least fairly present. In contrast, the comparison between categories mentioned only rarely in the first round ($f\leq5\%$) and categories with low practice values in the second round shows that 6 out of the 9 rarely mentioned categories in the first round fall below an average of 3.3 in the practice assessments of the second round and are thus considered as not very present in science education.

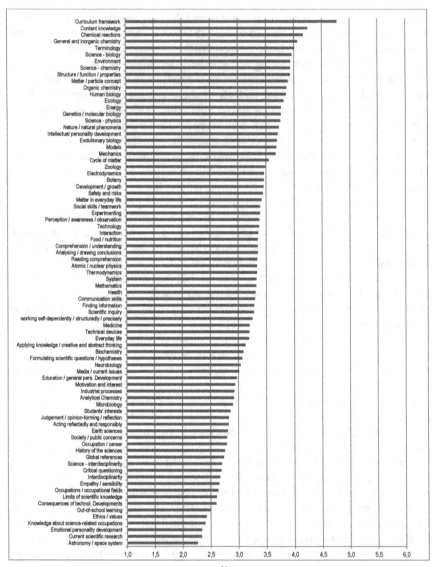

Figure 15. Practice assessments by the total sample[24]

24 This figure can also be accessed via www.springer.com and "Theresa Schulte" within the OnlinePLUS programme.

4.2.3.2   Sub-Samples

Additional insights into the assessment of the given aspects' realization in science education beyond the analysis of the practice values by the total sample can be gained by taking into account the different sub-samples. A first approach towards a differentiated view on the sub-samples is provided through their average practice assessments of all categories (Figure 16).

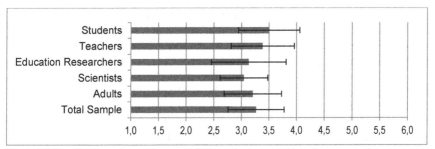

Figure 16. Overall mean values of the categories' practice assessments in the sub-samples

It can be seen that all mean values range between 3 and 3.5, indicating that in contrast to the priority assessments, all sub-samples assess the categories on average as being present to a rather low extent. The highest practice values are given by the group of students (M=3.5), followed by teachers (M=3.4) and science education researchers (M=3.1). The lowest practice values are provided by scientists (M=3.0). A comparison of the students and the group of adult sub-samples (M=3.2) shows a tendency towards slightly lower overall practice assessments in the group of adult sub-samples. The degree of differentiation when assessing the categories' realization in practice is indicated by standard deviations, which are represented by error bars in Figure 16. The highest standard deviation appears in the group of education researchers (SD=0.68). This shows that participants in this group differentiate to a higher extent between their assessments than the other groups. The lowest degree of differentiation in the practice assessments can be found in the group of scientists (SD=0.43). With a difference of 0.25 between the highest and lowest standard deviation, the groups' degrees of differentiation are, while being larger than in their priority assessments, still fairly similar.

In the same way as for the analysis of the priority assessments (cf. 4.2.2.2), further, more detailed comparisons of the practice assessments of the different sub-samples are carried out on the basis of pair comparisons among the sample groups regarding the mean values of the individual categories assessments. The results are displayed in Table 15, Table 16, Table 17 and Table 18, structured according to the different parts of the category system (contexts motives and situations, basic concepts and topics, fields and perspectives, and qualifications).

Although the different sub-samples mostly consider the same aspects as most and least important and thus generally resemble each other in their practice assessments, the results of the significance test (Mann-Whitney-U-Test) indicates that the sub-samples differ in a number of pair comparisons from each other.

All in all, a number of 138 statistically significant differences out of 480 possible pair comparisons among the four sub-samples can be identified. This corresponds to 29% of all pair comparisons and is slightly higher than in the priority assessments (24%)

The highest number of significant differences (N=34) can be found between teachers and scientists (43%), followed by a comparison of students and scientists (N=33), which correspond to 41% of these pair comparisons. Between students and education researchers, 28 significant differences in the practice assessments (35% of these pair comparisons) can be found. The comparison between education researchers and scientists reveals 18 significant differences (23% of these pair comparisons). On a similar level, the comparison between students and teachers shows 15 significant differences (19% of all cases in this group comparison. With 10 differences, the lowest number of significant difference can be found between teachers and education researchers (13% of these pair comparisons). In a comparison between the students and the group of adult sub-samples, 30 significant differences can be found (38% of the cases in this group comparison). As in the priority assessments, most significant differences between the sub-samples' assessments of the realization of the aspects appear within the part "qualification" (N=49).

As described in 4.2.2.2, the results of the total sample are unproportionally influenced by the sub-samples' assessments due to the sample distribution in the second round not being equally balanced with respect to the four sub-samples. Therefore, the results are normalized to obtain average values that are in equal parts composed of the average values of each of the four sub-samples. A comparison of the normalized values against the original values of the total sample reveals that no deviation surpasses an absolute value of $\Delta T=0.06$. In 70 out of 80 cases (88%), the deviations fall below $\Delta T=0.04$ and in 61 out of 80 cases even below absolute values of $\Delta T=0.03$ (76%). Relating these deviations to the comparison of original and normalized priority values, slightly higher differences between original and normalized values can be identified in the majority of practice assessments (see appendix).

Table 15
Significant Differences (Mann-Whitney-U-Test) and Mean Values of the Practice Assessments Concerning Situations, Contexts and Motives

Category	Significant Differences							Mean Values				
	S/T	S/E	S/Sc	T/E	T/Sc	E/Sc	S/A	S	T	E	Sc	Total
Education / general pers. development								3.0	3.0	3.0	2.8	3.0
Emotional pers. development								2.4	2.2	2.2	2.5	2.3
Intellectual personality development								3.8	3.8	3.7	3.5	3.7
Students' interests	*		*				*	3.2	2.7	2.9	2.8	2.9
Curriculum framework								4.7	4.8	5.0	4.7	4.8
Out-of-school learning								2.5	2.4	2.5	2.6	2.5
Media / current issues				*				2.9	3.2	2.8	3.0	3.0
Nature / natural phenomena	*						*	3.3	4.0	3.7	3.8	3.7
Medicine								3.0	3.5	3.0	3.2	3.2
Technology								3.2	3.5	3.6	3.2	3.4
Everyday life					*			3.0	3.5	3.2	2.9	3.2
Society / public concerns								2.9	2.9	2.6	2.8	2.8
Global references		*	*				*	3.2	2.8	2.3	2.7	2.7
Occupation / career	*	*					*	3.3	2.6	2.4	2.9	2.8
Science - biology			*			*		4.2	4.1	4.1	3.6	4.0
Science - chemistry			*	*	*	*	*	4.5	4.0	4.1	3.4	3.9
Science - physics			*			*		4.0	3.8	3.8	3.4	3.8
Science - interdisciplinarity		*	*				*	3.0	2.9	2.4	2.4	2.7
Number of stat. sig. differences	3	3	5	2	2	3	6					
				18								

Note. S=students, T=teachers, E=education researchers, Sc=scientists, A=adults (teachers, education researchers, and scientists), *=significant (p<0.05)

Table 16
Significant Differences (Mann-Whitney-U-Test) and Mean Values of the Practice Assessments Concerning Basic Concepts and Topics

Category	Significant Differences							Mean Values				
	S/T	S/E	S/Sc	T/E	T/Sc	E/Sc	S/A	S	T	E	Sc	Total
Matter / particle concept			*		*	*		4.2	4.1	4.0	3.4	3.9
Structure / function / properties			*		*		*	4.3	4.4	3.8	3.2	3.9
Chemical reactions			*		*	*		4.4	4.4	4.4	3.6	4.2
Energy					*			3.9	3.9	3.8	3.4	3.8
Development / growth			*		*		*	3.8	3.6	3.4	3.0	3.5
Models		*	*		*	*	*	4.4	4.0	3.6	2.9	3.7
Terminology	*	*	*		*	*	*	4.8	4.0	4.2	3.3	4.0
System			*		*			3.6	3.6	3.2	2.8	3.3
Interaction			*		*	*		3.6	3.8	3.3	2.7	3.4
Scientific inquiry	*	*	*		*		*	4.0	3.5	3.1	2.7	3.3
Limits of scientific knowledge		*	*				*	3.1	2.6	2.3	2.4	2.6
Cycle of matter			*		*			3.8	3.8	3.4	3.2	3.6
Food / nutrition	*				*			3.1	3.7	3.2	3.2	3.4
Health								3.3	3.4	3.1	3.3	3.3
Matter in everyday life					*			3.4	3.8	3.4	3.0	3.4
Technical devices		*					*	3.6	3.1	2.9	3.2	3.2
Environment								4.0	4.0	3.8	4.0	3.9
Industrial processes				*	*			3.0	3.3	2.7	2.6	2.9
Safety and risks	*	*	*				*	4.0	3.5	3.0	3.2	3.4
Occupations / occupational fields		*			*	*		3.1	2.4	2.1	2.9	2.6
Number of stat. sig. differences	4	7	12	1	15	6	8					
			45									

Note. S=students, T=teachers, E=education researchers, Sc=scientists, A=adults (teachers, education researchers, and scientists), *=significant ($p<0.05$)

Table 17
Significant Differences (Mann-Whitney-U-Test) and Mean Values of the Practice Assessments Concerning Fields and Perspectives

Category	Significant Differences							Mean Values				
	S/T	S/E	S/Sc	T/E	T/Sc	E/Sc	S/A	S	T	E	Sc	Total
Botany								3.3	3.4	3.6	3.5	3.5
Zoology		*					*	3.1	3.5	3.9	3.5	3.5
Human biology			*	*	*			4.1	3.8	4.3	3.4	3.9
Neurobiology				*				2.8	3.4	3.1	2.7	3.0
Genetics / molecular biology	*	*		*			*	4.4	3.9	3.7	3.2	3.8
Microbiology								3.1	2.9	2.9	2.8	2.9
Evolutionary biology				*				3.9	3.9	3.5	3.4	3.7
Ecology	*			*				3.6	4.2	3.7	3.6	3.8
General and inorganic ch.				*				4.0	4.3	4.2	3.7	4.1
Organic ch.				*	*			3.8	4.3	3.7	3.6	3.9
Biochemistry		*		*				3.3	3.3	2.7	2.9	3.1
Analytical Chemistry	*	*	*	*			*	3.4	3.2	2.3	2.7	2.9
Thermodynamics								3.6	3.6	3.0	3.1	3.3
Electrodynamics								3.5	3.6	3.3	3.3	3.5
Mechanics								3.4	3.5	4.1	3.7	3.7
Atomic / nuclear physics								3.6	3.3	3.3	3.2	3.3
Astronomy / space system						*		2.2	2.3	2.0	2.5	2.3
Earth sciences								3.0	2.7	2.5	3.0	2.8
Mathematics			*				*	3.8	3.4	3.3	2.9	3.3
Interdisc.								3.0	2.7	2.4	2.5	2.7
Current scientific research								2.6	2.4	2.0	2.3	2.3
Consequences of technol. Developments								2.9	2.6	2.3	2.6	2.6
History of the sciences		*		*				2.9	3.1	2.2	2.6	2.8
Ethics / values		*					*	2.6	2.4	2.0	2.6	2.4
Number of stat. sig. differences	1	6	4	4	8	3	4					

26

Note. S=students, T=teachers, E=education researchers, Sc=scientists, A=adults (teachers, education researchers, and scientists), *=significant ($p<0.05$)

Table 18
Significance Values (Mann-Whitney-U-Test) and Mean Values of the Practice Assessments Concerning Qualifications and Attitudes

Category	Significant Differences							Mean Values				
	S/T	S/E	S/Sc	T/E	T/Sc	E/Sc	S/A	S	T	E	Sc	Total
Motivation and interest			*		*			3.2	3.1	3.0	2.6	2.9
Critical quest.		*	*	*			*	3.1	2.9	2.2	2.5	2.7
Acting reflectedly and resp.		*					*	3.3	2.8	2.5	2.7	2.8
Knowledge about science-related occ.					*	*		2.7	2.3	2.0	2.6	2.4
Content knowl.					*	*		4.2	4.5	4.6	3.7	4.3
Comprehension / understanding			*		*	*	*	3.7	3.5	3.5	2.9	3.4
Applying knowledge/ creative/abstr. thinking		*	*		*		*	3.7	3.3	2.9	2.7	3.1
Judgement / opinion-forming / refl.	*	*	*	*			*	3.5	3.0	2.3	2.6	2.8
working self-dep./ structuredly / prec.	*	*	*				*	4.1	3.3	2.8	2.9	3.3
Finding inf.	*	*	*				*	4.0	3.3	3.0	3.0	3.3
Reading compr.	*	*				*	*	3.8	3.2	2.9	3.5	3.3
Comm. Skills								3.4	3.4	3.0	3.3	3.3
Social skills / teamwork	*	*	*				*	4.0	3.4	3.1	3.2	3.4
Empathy / sensibility		*		*		*		2.9	2.8	2.1	2.7	2.7
Perception / awareness / observation			*		*	*		3.5	3.6	3.4	3.0	3.4
Formulating sc. questions / hypotheses		*	*		*		*	3.6	3.3	2.9	2.5	3.1
Experiment.	*	*	*		*		*	4.1	3.5	3.3	2.9	3.4
Analysing / drawing concl.	*	*	*		*		*	4.0	3.4	3.2	2.9	3.4
Number of stat. sig. differences	7	12	12	3	9	6	12					

49

Note. S=students, T=teachers, E=education researchers, Sc=scientists, A=adults (teachers, education researchers, and scientists), *=significant ($p<0.05$)

4.2.4   Priority-Practice Differences

After analyzing the views of different stakeholders on the importance and degree of realization of aspects of desirable science education, the following section addresses a comparison of these two assessments to gain insights into areas of science education that feature most dissatisfaction and which are thus are in most urgent need for improvement. These insights are gained through an analysis of priority-practice differences. The priority-practice differences are calculated by subtracting the practice values from priority values, which are obtained from the assessments discussed in 4.2.2 and 4.2.3. In case of a positive difference value, the corresponding category is seen as underrepresented in practice, which means less realized than its priority assessment requires. In contrast, a negative difference value points to an overrepresentation of the respective category in practice. Large differences between priority and practice point to aspects that are considered relatively important but rarely realized, therefore indicating great dissatisfaction with the current state of science education and thus pointing to areas that need improvement. In contrast, small values represent sufficient correspondence between a category's realization in practice and its assigned importance.

4.2.4.1   Total Sample

A comparison of the priority and practice assessments by the total sample on the basis of average priority-practice differences of each category provides insights into aspects of desirable science education with most and least dissatisfaction in terms of their realization and thus areas that appear to be in most and least need for improvement. Part IV (methodical aspects) of the category system is not included in the comprehensive consideration of this comparison, as this section features irregular and incomplete data. The results are displayed in Figure 17 in descending priority-practice differences[25]. In general, all categories except "curriculum framework" feature positive priority-practice differences. In view of the 6-tier rating scales for the assessments of the categories' priority and realization, the mean of all priority-practice differences in the total sample (M=1.2) represents a rather large discrepancy between desirable science education and its realization. More specifically, 65 out of the 80 given categories feature priority-

---

25 In order to display the spectrum of the differences in descending order according to the extent of their intervals between priority and practice values without taking into account over or underrepresentation, the priority-practice difference of "curriculum framework" (the only aspect with a negative priority-practice difference) is for representation within this diagram transformed into an absolute value.

practice differences with mean values of M>0.75, which equals 81% of all categories. "Curriculum framework" (M= -0.9) is the only category with negative difference entries and thus with an overall negative difference value, indicating an observed overrepresentation of this aspect in science education by all participants of this study.

As very high or very low difference values inform about especially big or small discrepancies between priority and reality, special emphasis is placed on addressing aspects with extreme differences between their given priority and their perceived realization in practice. The largest priority-practice differences can be found for "critical questioning" (M=2.4), "judgement / opinion-forming / reflection" (M=2.3), "acting reflectedly and responsibly" (M=2.2), "current scientific research" (M=2.1), "motivation and interest" (M=2.1), "students' interests" (M=2.1), "ethics / values" (M=2.0), "consequences of technological developments" (M=2.0), "applying knowledge / creative and abstract thinking" (M=2.0), and "interdisciplinarity" (M=1.9). All of these aspects can be found among those categories that are in the priority assessments given values greater or equal to the mean value of all priority assessments (M=4.4). Therefore, the highest discrepancies appear for those areas that are at the same time considered as most important. The smallest differences can be identified for "zoology" (M=0.2), "chemical reactions (M=0.2), "general and inorganic chemistry (M=0.3), "botany" (M=0.3), "terminology" (M=0.3), and "evolutionary biology (M=0.3). This shows that the relation between the priorities of these categories and their presence in the science classroom is seen by the stakeholders as most balanced and thus considered in their realization as closest to their priorities.

With the standard deviation as an indication for the degree of conformance among the priority-practice differences, further insights can be gained whether a category shows similar gaps between its priority and realization, or if the gaps between priority and realization for a category differ. In this way, the standard deviation provides hints about which aspects share most agreement in their priority-practice differences. The highest standard deviations and therefore lowest degree of consensus appears for "ethics / values" (SD=1.72), "mathematics" (SD=1.62), "botany" (SD=1.62), "science – interdisciplinarity" (SD=1.59), "limits of scientific knowledge" (SD=1.59), "models" (SD=1.58), "empathy / sensibility" (SD=1.57), "interdisciplinarity" (SD=1.55), "history of the sciences" (SD=1.55), and "finding information" (SD=1.55). The lowest standard deviations and thus highest degree of consensus can be found for "general and inorganic chemistry" (SD=1.10) and "organic chemistry" (SD=1.10).

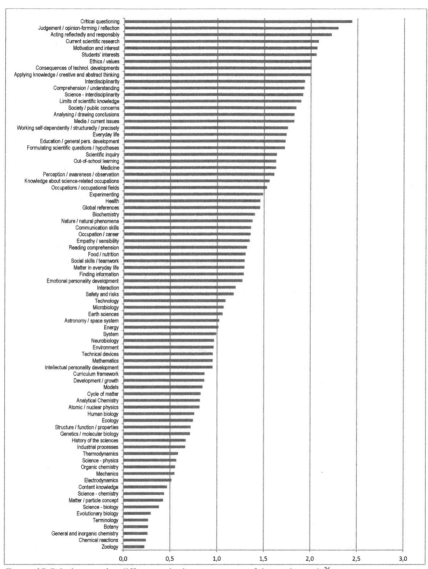

Figure 17. Priority-practice differences in the assessments of the total sample[26]

---

26 This figure can also be accessed via www.springer.com and "Theresa Schulte" within the OnlinePLUS programme.

A comparison of the standard deviations of the ten categories with the highest priority-practice differences against the mean value of all standard deviations (M=1.39) shows that seven out of these ten categories have standard deviations that are smaller or equal to this value and thus feature comparably strong consensus. For the ten categories with lowest priority-practice differences, neither particularly strong nor particularly weak consensuses in terms of dissatisfaction can be found.

4.2.4.2 Sub-Samples

In addition to the analysis of the priority-practice differences in the total sample, a more differentiated view about what is seen to be important and what is perceived to determine present science education is provided by a consideration of the priority-practice differences in the sub-samples. A first approach towards a more differentiated analysis is provided through the average of all priority-practice differences in the four sub-samples[27] (Figure 18). This analysis reveals in which sub-samples the given priorities of aspects of desirable science education and their realization in the classroom feature the largest gap and thus which sub-samples consider science education as most deficient with respect to fulfilling their expectations.

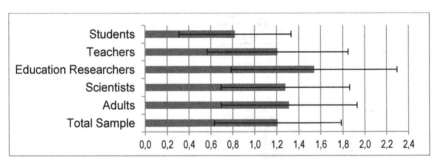

Figure 18. Overall mean values of the categories' priority-practice differences in the sub-samples

27 As the focus of this representation are the scopes of the differences in the sub-sample groups, absolute values of the sub-samples' average priority-practice differences are employed for this bar chart. In addition to "curriculum framework" in all sub-samples, negative values only occur for "chemical reactions" (M= -0.1), "models" (M= -0.2), and "terminology" (M= -0.3) in the group of students and for "zoology" (M= -0.2) in the group of scientists. This indicates that in one out of four sub-samples, participants considered the aforementioned categories as slightly overrepresented, whereas the majority of stakeholders perceive them slightly underrepresented.

It can be seen that all sub-samples feature priority-practice differences between 0.8 and 1.5, thus showing general dissatisfaction with the current state of science education compared with the given priorities. The highest average divergence between priority and realization can be found in the group of science education researchers (M=1.5). This means that the science education researchers express the highest overall dissatisfaction with the categories' representations in practice. In contrast, the smallest gap between priority and practice appears in the group of students (M=0.8), indicating that the students consider all given aspects on average as comparably more adequately represented in the classroom than the other sub-samples and thus express lowest dissatisfaction with current science education. Comparing the students with the group of adult sub-samples (M=1.3), higher dissatisfaction in the group of adult sub-samples can be noted. These trends are also underlined by a consideration of the values of the individual priority-practice differences. While in the group of students, 46 out of 80 priority-practice differences (58%) show mean values of M>0.75, this is the case for teachers in 63 out of 80 differences (79%), in the group of scientists in 68 out of 80 differences (85%) and in the group of education researchers in 71 out of 80 differences (89%). In the adult sub-samples altogether, 69 out of 80 differences (86%) with M>0.75 can be found.

The standard deviations, as indicated by error bars in Figure 18, illustrate the scope of the priority-practice differences in the different sub-samples. The highest standard deviation appears in the group of education researchers (SD=0.75), which indicates that in this group, the degrees of dissatisfaction with the priority-practice relation of the individual aspects deviate most. The lowest scope in priority-practice differences can be found in the group of students (SD=0.51), which shows that the degrees of dissatisfaction with the realization of the individual categories in practice are more similar in the group of students than in the other sub-samples.

In order to gain more detailed insights into the dissatisfaction of the different groups regarding the given aspects' relation between priority and their realization, further comparisons are made on the basis of the results displayed in Table 19, Table 20, Table 21,and Table 22. The tables represent the different sections of the category system (contexts motives and situations, basic concepts and topics, fields and perspectives, and qualifications).

It can be seen that in general, the different sub-samples consider mostly the same aspects as most and least deficient with regard to their priority-practice relation. Yet, the results of the significance test (Mann-Whitney-U-Test) indicate that the sub-samples differ from each other in several pair comparisons. In total, a number of 130 statistically significant differences out of 480 possible pair comparisons among the four sub-samples can be found. This corresponds to 27%

of the cases of all pair comparisons and is similar to the percentages of significant differences in the priority (24%) and practice (29%) assessments.

Again, the highest number of significant differences (N=39) appears between the students and education researchers (49% of all cases of this pair comparison), followed by students and teachers (N=27, 34% of all cases of this pair comparison), and students and scientists (N= 25, 31% of all cases of this pair comparison). Between education researchers and scientists, 17 significant differences can be identified (21% of all cases of this pair comparison). The lowest number of significant differences can be found between teachers and education researchers (N=12, 15% of all cases of this pair comparison) as well as between teachers and scientists (N=10, 13% of all cases of this pair comparison). When contrasting the students with the group of adult sub-samples, 29 significant differences (36% of the cases in this group comparison) appear. As expectable on the basis of the findings from the differentiated analyses of the priority and practice assessments, most of the significant differences between the priority-practice differences of the sub-samples appear in the part "qualification" (N=34).

In order to account for the unbalanced composition of the sample, a comparison of normalized values with the original values is carried out. An analysis of this comparison shows that all discrepancies between original and normalized priority-practice differences range at absolute values of $\Delta T=0.04$ or below. In 46 out of 80 cases (75%), the differences even fall below 0.03 (see appendix). These values are similar to those of the priority and practice differences.

Table 19

Significance Values (Mann-Whitney-U-Test) and Mean Values of the Priority-Practice Differences Concerning Situations, Contexts and Motives

Category	Significant Differences S/T	S/E	S/Sc	T/E	T/Sc	E/Sc	S/A	Mean Values S	T	E	Sc	Total
Education / gen. pers. developm.	*	*					*	1.4	1.7	2.1	1.8	1.7
Emotional pers. development	*	*					*	0.8	1.6	1.6	1.0	1.3
Intellectual pers. developm.								0.7	0.9	1.1	1.1	1.0
Students' int.	*	*						1.7	2.3	2.4	1.8	2.1
Curric. framew.								-0.8	-0.8	-1.1	-0.9	-0.9
Out-of-school learning	*							1.1	2.0	1.7	1.6	1.6
Media / current issues.								1.8	1.8	2.0	1.8	1.8
Nature / natural phenomena								1.3	1.3	1.7	1.4	1.4
Medicine								1.7	1.4	1.8	1.6	1.6
Technology								0.8	1.1	1.0	1.4	1.1
Everyday life		*						1.5	1.7	2.2	1.7	1.7
Society / public concerns		*				*		1.5	2.0	2.3	1.5	1.8
Global ref.		*				*		1.1	1.6	2.1	1.1	1.5
Occ./ career								1.0	1.6	1.8	1.1	1.4
Science - biol.					*	*		0.5	0.1	0.2	0.8	0.4
Science – ch.			*		*	*		0.0	0.3	0.3	1.1	0.4
Science - phys.			*		*	*	*	0.0	0.4	0.5	1.3	0.6
Sc. - interdisc.	*	*	*				*	1.2	1.9	2.2	2.3	1.9
Number of stat. sig. differences	4	7	4	0	3	5	4		23			

Note. S=students, T=teachers, E=education researchers, Sc=scientists, A=adults (teachers, education researchers, and scientists), *=significant (p<0.05)

Table 20
Significance Values (Mann-Whitney-U-Test) and Mean Values of the Priority-Practice Differences Concerning Basic Concepts and Topics

Category	Significant Differences							Mean Values				
	S/T	S/E	S/Sc	T/E	T/Sc	E/Sc	S/A	S	T	E	Sc	Total
Matter / particle concept								0.0	0.5	0.6	0.6	0.4
Structure / function / properties		*	*		*		*	0.1	0.6	0.9	1.2	0.7
Ch. reactions			*					-0.1	0.3	0.1	0.5	0.2
Energy			*					0.6	1.0	1.1	1.3	1.0
Development / growth								0.4	0.9	1.0	1.1	0.9
Models	*	*	*				*	-0.2	0.9	1.3	1.3	0.8
Terminology	*			*	*		*	-0.3	0.8	0.1	0.1	0.3
System			*				*	0.5	0.8	1.2	1.4	1.0
Interaction	*	*	*	*			*	0.4	1.0	1.3	1.9	1.2
Scientific inquiry	*	*	*				*	0.6	1.6	2.2	2.1	1.6
Limits of sc. knowledge	*	*				*	*	1.1	2.0	2.7	1.8	1.9
Cycle of matter	*	*	*				*	0.1	0.9	1.2	1.1	0.8
Food / nutrition								1.2	1.2	1.6	1.3	1.3
Health								1.5	1.3	1.7	1.5	1.5
Matter in everyday life								0.9	1.3	1.6	1.5	1.3
Technical devices			*		*			0.5	0.7	1.6	1.2	1.0
Environment								0.9	1.0	1.3	0.8	1.0
Industrial processes						*		0.6	0.2	0.9	1.1	0.7
Safety and risks	*	*					*	0.5	1.3	1.5	1.3	1.2
Occupations / occ. Fields	*	*					*	1.1	1.8	2.2	1.1	1.5
Number of stat. sig. differences	8	9	8	2	4	2	9					
			33									

Note. S=students, T=teachers, E=education researchers, Sc=scientists, A=adults (teachers, education researchers, and scientists), *=significant (p<0.05)

Table 21
Significance Values (Mann-Whitney-U-Test) and Mean Values of the Priority-Practice Differences Concerning Fields and Perspectives

Category	Significant Differences							Mean Values				
	S/T	S/E	S/Sc	T/E	T/Sc	E/Sc	S/A	S	T	E	Sc	Total
Botany								0.0	0.4	0.8	0.0	0.3
Zoology						*		0.3	0.4	0.5	-0.2	0.2
Human biology								0.5	0.9	0.8	0.8	0.8
Neurobiology	*							1.5	0.7	1.0	0.8	1.0
Genetics / molecular biology			*					0.3	0.6	0.8	1.1	0.7
Microbiology								0.9	1.0	1.2	1.2	1.1
Evolutionary biol.								0.1	0.1	0.8	0.3	0.3
Ecology		*		*				0.5	0.6	1.3	0.7	0.7
General and inorganic chemistry								0.4	0.2	0.2	0.3	0.3
Organic chemistry		‚						0.9	0.5	0.5	0.5	0.6
Biochemistry								1.2	1.3	1.9	1.4	1.4
Analytical Chemistry		*		*				0.7	0.4	1.5	0.9	0.8
Thermodynamics		*						0.1	0.5	1.0	0.7	0.6
Electrodynamics								0.3	0.4	0.8	0.6	0.5
Mechanics								0.6	0.6	0.3	0.6	0.5
Atomic / nuclear physics								0.9	0.8	1.1	0.6	0.8
Astronomy / space system		*		*		*		0.8	0.9	1.8	0.8	1.0
Earth sciences		*		*		*		0.7	1.0	1.8	0.8	1.1
Mathematics			*			*		0.5	0.9	0.7	1.6	1.0
Interdisciplinarity	*	*	*				*	1.0	2.1	2.5	2.0	1.9
Current scientific research								2.2	2.2	2.4	1.7	2.1
Consequences of technol. Developments		*		*		*		1.8	2.0	2.7	1.6	2.0
History of the sciences				*		*		0.6	0.4	1.4	0.6	0.7
Ethics / values	*	*					*	1.2	2.2	2.7	1.8	2.0
Number of stat. sig. differences	3	8	3	6	0	6	2					

26

Note. S=students, T=teachers, E=education researchers, Sc=scientists, A=adults (teachers, education researchers, and scientists), *=significant (p<0.05)

Table 22
Significance Values (Mann-Whitney-U-Test) and Mean Values of the Priority-Practice Differences Concerning Qualifications and Attitudes

Category	Significant Differences							Mean Values				
	S/T	S/E	S/Sc	T/E	T/Sc	E/Sc	S/A	S	T	E	Sc	Total
Motivation and interest								1.8	2.0	2.4	2.2	2.1
Critical questio-nung		*	*	*			*	1.8	2.4	3.0	2.6	2.4
Acting reflectedly and responsibly	*	*				*	*	1.4	2.5	2.9	2.0	2.2
Knowledge about science-related occupations	*	*			*	*		1.1	1.8	2.2	1.2	1.6
Content knowl.								0.5	0.3	0.3	0.6	0.5
Comprehension / understanding		*	*		*		*	1.3	1.8	2.2	2.3	1.9
Applying knowledge / creative and abstr. thinking	*	*	*				*	1.1	2.0	2.5	2.3	2.0
Judgement / opinion-forming / reflection	*	*	*	*			*	1.3	2.3	3.0	2.5	2.3
working self-dependently / structuredly / precisely	*	*	*	*			*	0.8	1.8	2.4	1.9	1.8
Finding inform.	*	*	*				*	0.0	1.5	1.9	1.4	1.3
Reading compr.	*	*	*		*	*	*	0.2	1.8	2.0	1.0	1.3
Comm. Skills	*	*					*	0.8	1.4	1.8	1.3	1.4
Social skills / teamwork	*	*					*	0.7	1.6	1.8	1.0	1.3
Empathy / sens.		*		*		*	*	0.8	1.4	2.3	1.0	1.3
Perception/ awareness/ obs.								1.1	1.6	1.8	1.8	1.6
Formulating sc. questions/hyp.	*	*	*				*	0.9	1.6	2.3	2.1	1.7
Experimenting	*	*	*				*	0.7	1.6	1.8	1.7	1.5
Analysing / drawing concl.	*	*	*				*	0.8	1.8	2.2	2.3	1.8
Number of stat. sig. differences	12	15	10	4	3	4	14					
				48								

Note. S=students, T=teachers, E=education researchers, Sc=scientists, A=adults (teachers, education researchers, and scientists), *=significant (p<0.05)

4.2.5   Discussion

In the second round, 154 out of the contacted 193 stakeholders from the first round returned for participation. With a response rate of 80%, this remaining sample for the second round is very satisfying for a multi-round and long-term study such as the Berlin Curricular Delphi Study in Science. The discussion of the findings of the first part of the second round in the following sections is structured according to the three emphases of the analysis in terms of priority, practice, and priority-practice differences.

Priority Assessment

The results of the analysis of the priority assessments yield insights into the importance which different stakeholders in society that are affected by science education assign to various aspects of science education. A comparison of the different priority values of the categories shows that all categories but one reach average priorities above the theoretical mean values of M=3.5, which shows that the categories derived from the participants' statements in the first round are considered as important and relevant in the second round. This can be seen as a confirmation of the validity of the classification system. As such confirmation can only be reached through iteration – one of the central features of the Delphi method – this is one of the findings that particularly result from the strength of the applied method.

In the curricular Delphi study in chemistry (Bolte, 2003a), particular importance is given to aspects that can be assigned to more overarching aims of education as well as to everyday life related aspects (cf. 2.4.3). This result is similar to the findings of the curricular Delphi study in physics by Häußler (1980), which include among highly prioritized aspects topics related to "scientific knowledge and methods as mental tools", "physics as a vehicle to promote practical competence", and "physics as a socio-economic enterprise", showing that the participants see physics "more as a human enterprise and less as a body of knowledge and procedures" (Häußler & Hoffmann, 2000, p. 704).

The most commonly recommended context-based issues in European science education are related to contemporary societal issues. Moreover, environmental concerns and the application of scientific achievements to everyday life are recommended for discussion in science lessons (Eurydice, 2011, p. 85). For students as one particular stakeholder group, PISA Science (OECD, 2007b) shows that they are most interested in learning about health or safety issues that they might encounter personally, and were least interested in topics that they see as having little personal relevance. This is found to be in agreement with their clear preference for human biology as the broad science area that they had most interest in learning about (Bybee & McCrae, 2011, pp. 23-25).

In the opinion of the sample participating in the Berlin curricular Delphi study in Science, similar general tendencies in the priorities can be identified. However, contrary to previously made assumptions, aspects related to interdisciplinarity, the relation between science and everyday life, society, and media do not feature the strongest emphases but can be found in the upper mid-range. Instead, the priorities of the stakeholders in the Berlin Curricular Delphi Study in Science mostly correspond to the broad spectrum of both general and scientific inquiry related skills of scientific literacy. Similar to the findings by Bolte (2003a), aspects considered by the participants in the second round of this study as most important refer to general skills and overarching competences such as "comprehension / understanding", "analysing / drawing conclusions", "applying knowledge / creative and abstract thinking", "judgement / opinion-forming / reflection", "critical questioning", "motivation and interest" and "acting reflectedly and responsibly", as well as to elements that can be assigned to scientific inquiry such as "working self-dependently / structuredly / precisely" and "perception / awareness / observation". Therefore, hypothesis 2a, assuming that highest priority is given to aspects related to scientific inquiry, environmental issues, content knowledge, and overarching aims of general education can only partly be confirmed. Yet, although "environment" and "content knowledge" are not in the top ten categories, these categories are assessed with mean values of M>4.5 and thus still range among 20% of the most important categories.

Relating the most highly prioritized categories to those aspects mentioned by most stakeholders in the first round, it can be seen that except one, all categories referenced by more than 25% of the participants in the first round reach in the second round mean values above the mean value of all priority assessments (M=4.4). In this way, the most frequently mentioned categories in the first round are endorsed in the second round in terms of their general priority. This further proves the validity of the classification system. However, the most frequently mentioned aspects in the first round ("everyday life" and "media / current issues"), although in the second round still assessed comparatively high, do not appear among the ten highest priority assessments and thus do not receive the same emphasis as in the first round. This suggests that aspects related to students' everyday life and to current media coverage surely represent important parts of scientific literacy based science education, but aspects such as certain general and scientific competences are considered more important by the stakeholders.

Several additional categories, which were not mentioned by a high percentage of participants in the first round and thus possibly considered as being not so relevant, are in the second round indeed considered as highly important for science education. This implies that in light of the results of the second round, the frequency results of the first round have to be relativized. However, this does not make the results of the first round unreliable. Rather, these shifts in the specific accentuations of the stakeholders between the first and second round show that no preliminary conclusions should be drawn from such first findings. Instead, the more specific practice assessments in the second round show that solid assessments on complex issues such as desirable aspects of science education cannot be gained through single surveys, but rather require several iterations, as facilitated by the Delphi method.

Similar to the results of Bolte (2003a), lower values, yet still at least fairly important priority values are assigned to categories that relate to more specific scientific perspectives as well as to scientific sub-disciplines and their traditional division. Hence, hypothesis 2b, stating that lowest priority is given to aspects connected to the structure of the science disciplines, specialized fields, and traditional approaches of single-subject-orientation can be confirmed. Most of these lower priorities are assigned to those categories rarely mentioned in the first round. This shows that although many of these categories were only rarely referred to in the first round, the participants in fact still consider these categories in view of all categories as at least fairly important for scientific literacy based science education. Again, this result reveals that the frequencies of the first round are not entirely reflected in the priority assessments of the second round and that thus, an application of the Delphi method that allows the participants to reconsider their opinion in view of all categories over the rounds has proven useful. In this way, these findings point to one of the main strengths of the Delphi methods of gaining more solid results through iteration.

A consideration of the standards deviations of the individual categories serves as a first approach towards a more differentiated analysis of the priority assessments. With standard deviations between 0.8 and 1.3 and more than 50% of the categories not surpassing a standard deviation of 1.0, it can be said that the participants assess the categories at least on fairly similar levels. This implies a certain consensus among the stakeholders in their priority assessments and thus, generally similar opinions about aspects of desirable science education. Hence, hypothesis 5a is supported in the second round as well.

The highest standard deviations and thus lowest degree of consensus in the priority assessments among the participants appears for "ethics / values" (SD=1.3) and "emotional personality development (SD=1.3)". Reasons for this finding might be found in the rather broad range of meanings covered by these terms and thus possibly different associations.

When taking a closer look at the priority assessments of the four sub-samples in terms of their average assessments of all categories, the categories can be found to be assessed by all sub-samples on average with "rather high priority". The highest overall priority assessments are given by the science education researchers, which suggests that this group places higher demands towards science education. Similar implications can be found for the group of adult sub-samples with comparably high overall priority values as well. In contrast, with the lowest overall priorities, the students seem to share in general slightly lower demands towards science education. This might imply that several aspects of science education, while still considered as important, are not as important to the students as to the adult sub-samples and supports the assumption of a gap between the opinions of students and adult stakeholders (hypothesis 5c).

The highest degree of differentiation between the priorities of the categories is found in the group of teachers, which suggests that stakeholders in this group consider some categories as especially important, whereas other categories are deemed considerably less relevant. The lowest degree of differentiation in the priority assessments among the sub-samples appears in the group of students. Relating this finding to their overall priority assessments suggests that they consider all of the given categories on a rather similar and at the same time slightly lower level of importance. Yet, as the groups do not considerably differ in their degrees of distinction among the category priorities, this finding should be considered with reservation.

A closer look at the specific assessments of the stakeholders reveals a number of different accentuations[28] among the different sample groups in terms of what is seen to be important in science education. However, with a total of 117 statistically significant differences out of 480 possible pair comparisons among the four sub-samples (representing 24% of all pair comparisons), the hypothesis that the stakeholder groups differ in their specific assessments of desirable science education (hypothesis 5b) cannot fully be confirmed.

With only 7 significant differences (8% of these pair comparisons), the highest consensus in the priority assessments can be found between teachers and education researchers. This suggests a particular proximity of these two stakeholder groups regarding their priorities in meaningful science education. Reasons for this similarity might relate to the fact that with the first phase of teacher education in Germany taking place in education research departments at university, both groups share a background of theory-based science educational issues.

---

28   With regard to differences between sub-samples, it should be noted that the assessments of the sub-samples do not contradict each other in these cases but rather indicate significant differences and thus certain deviations within their emphases.

However, a verification of this assumption would require further investigations. The largest disagreement on the priority of the given categories (N=27) appears between students and education researchers and between teachers and scientists (34% of these pair comparisons). The contrast between students and science education researchers corresponds to their different response behaviors in the first round.

With the pair comparison between students and the group of adult sub-samples revealing 25 significant differences (31% of these comparisons), the hypothesis of a particular gap between the views of the young and adult generation on desirable science education can only partly be confirmed for the priorities in the second round. Although the majority of aspects are assessed with similar tendencies, the considerable number of differences in this pair comparison makes clear that in several aspects, students and adults in fact differ in their expectations. Thus, this finding at least questions the frequently invoked consensus about desirable aspects of science education. How far the opinions of these two groups converge throughout the further course of the study is shown by the results of the third round (see 4.3).

The fact that most significant differences (N=34) among the pair comparisons appear in the field of "qualification" shows that the sub-samples, while not contradicting each other, feature most differing accentuations and thus lowest consensus regarding the importance of categories relating to attitudes and competences that should be enhanced in science education. Reasons for this finding might relate to the stakeholders of the different sub-samples having dealt with competences from different perspective, e.g. the scientists from a more practical, work life view, the science education researchers from a more theory-based perspective, the teachers from a more school practical view, and the students, in contrast, from the most inexperienced perspective in terms of life skills and life-long learning. These differences might explain why the stakeholder groups especially deviate in their validation of different competences of desirable science education.

All in all, the fact that almost all categories are considered at least fairly important, and that the stakeholders assessed all categories on average with "rather high priority" shows that the categories developed in the first round of this study are suitable to represent important aspects in the views on desirable science education of different stakeholder groups in society that are affected by science education. The results of the differentiated analyses according to different sub-samples show that, despite several differences, the sub-samples mostly agree on aspects of desirable and contemporary scientific literacy based science education. The largest consensus appears between teachers and education researchers, while the group of students and adult sub-samples more frequently differ in their emphases.

To what extent these aspects are realized in current science education in Germany is shown by the practice values.

Practice Assessment

The results of the analysis of the practice assessments provide insights into the extent to which the aspects represented by the given categories are realized in the science classroom according to the stakeholders. In contrast to the priority assessments, in which all categories but one reach on average priorities above the theoretical mean values of 3.5, only 22 out of 80 categories (28%) are considered by the total sample as at least fairly present in science education. No category is perceived present "to a high extent". Instead, most aspects are assessed as realized to a rather low or low extent and thus seen as being not very present in science education.

A notable finding is that, in contrast to the priority assessments, mainly categories related to scientific concepts, the traditional division of the science subjects, content knowledge, and scientific disciplines as well as specialized subdisciplines can be found among the categories with the highest practice values. These emphases are as well reflected by "curriculum framework", the category perceived as most present in science education. Hence, hypothesis 3a, stating that aspects related to the structure of the science disciplines, specialized fields, and traditional approaches of single-subject-orientation are assessed with the highest extent of realization, can be confirmed. This result is very similar to the findings of the curricular Delphi study in chemistry, which reveal highest realization of aspects related to terminology, basic terms, content knowledge, and scientific concepts (Bolte, 2003a). Further comparisons can be drawn to a study by Osborne and Collins (2000), which describe science education in the opinion of students and parents as an experience dominated by scientific content and as a fragmented endeavor that leaves students without sufficient literacy.

A comparison of the ten highest priority values with the ten highest practice assessments shows that none of the priority top ten categories correspond to the practice top ten categories. This reveals a mismatch between the expectations of the stakeholders towards science education and how they perceive the reality. As more precise insights into possible mismatches between wish and reality are provided by the priority-practice differences (see 4.2.4), this finding is further discussed in that section.

The lowest degrees of realization mostly refer to scientific perspectives, such as "limits of scientific knowledge", "consequences of technological developments", "current scientific research", "ethics / values", to scientific sub-fields such as "astronomy / space system", and to specific aspects of science learning such as "out-of-school learning", "emotional personality development" and categories related science-related occupations. Although most of the aforementioned aspects are considered as at least fairly important facets of science education by the stakeholders, the practice assessments show that according to the stakeholders, current science education does not enhance the students' understanding about the nature of science, as it does not sufficiently address limits of scientific knowledge, positive and negative consequences of technological developments, and current scientific research. Also, science education is not seen to support students enough in gaining orientation about possible scientific occupations; neither is science education considered to contribute sufficiently to the students 'emotional personality development or their involvement with ethical issues in science.

In parts, these results are again similar to the findings in the curricular Delphi study in chemistry, which reveals lowest realization in practice for aspects related to overarching aims of general education and everyday life related aspects of science education (Bolte, 2003a). Yet, expect for ethical references, aspects of interdisciplinarity, students' living environment, and overarching aims of general education do not represent the lowest practice assessments in the findings of the second round of this study. Therefore, hypothesis 3b cannot fully be confirmed.

A comparison of the practice assessments of the second round to the most frequently mentioned categories in the first round ($f \geq 25\%$) reveals that only very few of the frequently mentioned aspects are considered as at least fairly present. On the other hand, 6 out of the 9 rarely mentioned categories ($f \leq 5\%$) in the first round fall below the theoretical mean value of $M=3.5$ in the practice assessments of the second round and are thus considered as not very present in science education.

One assumption derived from the category frequencies in the first round (cf. 4.1.4) following Bolte (2003a) is that comparatively low category frequencies might be related to the aspects subsumed by these categories playing a minor role in science education practice so that certain stakeholder groups such as students or scientists were not aware of these aspects in their open-text responses in the first round without exemplifying references and thus were not able to mention these categories. The analyses of the practice assessments support this assumption, as the results reveal that although the majority of the categories are considered in the second round as important by the stakeholders, only few of these aspects are seen to be comprehensively implemented or even sufficiently addressed and realized in practice. Thus, it is not surprising that certain aspects were in the first round not mentioned to the degree that would have been expected on the basis of current recommendations for modern scientific literacy based science education. Hence, explaining the cases of low category frequencies in the first round only with lacking reflection would be too short-sighted. Again, this finding points towards a strength of the Delphi method, as applying this method prevents short-sighted conclusions.

With the standard deviations of the individual categories as an indication for the degree of agreement among the participants in their assessments, there is moderate agreement among the participants with respect to what aspects they consider as most and least realized in current science education, as standard deviations of the different categories range between SD=0.9 and SD=1.3. In general, the degree of agreement among the participants about the realization of the given aspects (M_{SD}=1.1) is similar to degree of agreement among the participants about the priority of the given aspects (M_{SD}=1.0).

Further insights into the assessment of the given aspects' realization in science education are obtained from an analysis of the practice values of the different sub-samples. A closer look at the sub-samples' practice assessments of all categories reveals that, in contrast to the priority assessments, all sub-samples assess the categories as being present in science education to a rather low extent on average. The highest practice values are given by the group of students (M=3.5), which shows that this group perceives the aspects as realized to a slightly higher degree than the other groups. The lowest practice values (M=3.0) among all four sub-samples can be found in the group of scientists. Reasons for this finding might be related to this group's points of views being mostly influenced by work experiences or research perspectives and perceiving the level of scientific literacy of young people entering scientific apprenticeships, further training, programs at university etc.. However, further research would be necessary for a verification of this assumption. When contrasting the group of students and adult sub-samples, a tendency towards slightly lower overall practice assessments appears in the adult sub-samples. This finding might be explained by the young generation's experiences in science education being most immediate and thus resulting in perceptions of a higher presence of the given aspects.

The highest degree of differentiation in assessing the realization of categories appears in the group of education researchers, which implies that stakeholders in this group distinguish to a greater extent than the other groups between aspects they perceive as more and less realized in the science classroom. Relating this finding to their similarly high degree of differentiation in the priority assessments, a general attitude of assessing science education on a more differentiated level can be attributed to the science education researchers. The lowest degree of differentiation in the practice assessments among the sub-samples can be found in the group of scientists. A comparison of this finding to this group's general practice assessments implies that they see all given categories on a rather similar and at the same time slightly lower level of realization. In the assessments of the students, it is notable that while they place the lowest demands towards science education among all sub-samples, they perceive the given aspects at the same time as most realized among all sub-samples.

Although the different sub-samples mostly consider the same aspects as most and least realized, with a total of 138 statistically significant differences out of 480 possible pair comparisons (representing 29% of all pair comparisons), the sub-samples do not differ significantly in the majority of practice assessments. Thus, the assumption that the stakeholder groups differ in their specific assessments of desirable science education (hypothesis 5b) cannot fully be confirmed for the practice assessments either. With only 10 differences (13% of these pair comparisons), the highest consensus about the realization of the given aspects can be identified between teachers and education researchers, which again might be related to their professionally based similar perspectives on science education. However, further research would be needed for a verification of this assumption. Most disagreement about the realization of the given categories (N=34) can be found between teachers and scientists (43% of these pair comparisons), which points to rather large disagreement and reiterates the different response behaviors of these two groups in the first round. These findings might again be explained by the different professional backgrounds of the stakeholders in the different sub-samples.

With 30 significant differences in the pair comparison between students and the group of adult sub-samples (38% of the cases in this group comparison), different tendencies, but no overarching gap can be identified between the opinions of the young and adult generation concerning the current state of science education. Therefore, hypothesis 5c proposing a particular gap between the views of the generations can only partly be confirmed. However, the high number of differences in this pair comparison indicates that in several aspects, students and stakeholders from the adult sub-samples in fact differ in their perception of the realization. Thus, although no clear gap can be identified between the students and adult stakeholders, this finding might raise further questions with respect to the different perceptions about the current state of science education of students and adult stakeholders.

The fact that, in the same way as in the priority assessments, most significant differences (N=49) between the pair comparisons of the practice assessments appear for categories related to general and scientific competences shows that the sub-samples again feature most differing accentuations and thus lowest consensus in the field of "qualification". As suggested in 4.1.4, reasons for this finding might relate to differences in the sub-samples competence based perspectives due to their professional backgrounds.

In conclusion, the descriptive statistical results of the practice assessments show that only few aspects of desirable science education are considered as at least fairly present in the science classroom. Instead, most given categories are assessed as being realized in science education to a rather low or low extent and in this way seen as being not very present in science education. The average practice mean value of all categories (M=3.3) sharply contrasts with the average priority assessments of the categories (M=4.4). Also, when comparing the highest priority assessments and the highest practice assessments, it can be seen that the categories considered most important are in the opinion of the stakeholders not even realized to an at least fairly high extent. Moreover, the findings of the practice assessments illustrate that science education is seen to be mostly dominated by aspects related to the structure of the science disciplines and subject-specific fields, and traditional approaches of single-subject-orientation. In contrast, addressing issues that refer to aspects of more general education seems to play a minor role in practice. That subject-specific aspects essentially determine science education is not surprising; after all, science education is inherently about science. Yet, that the alignment to subject-specificity is so dominant that overarching aims of general education are given little attention is a notable finding. The results of the differentiated analyses according to the different sub-samples show that despite several significant differences, the sub-samples mostly agree on what determines current science education. The largest consensus about the realization of the given aspects can be found between teachers and education researchers, while the students and adult sub-samples partly differ in their emphases in the same way as in the priority assessments.

To what extent the priority and practice assessments of aspects of desirable science education relate to each other is shown in more detail on the basis of priority-practice differences.

Priority-Practice Differences

The difference between the priority and practice assessments are calculated by subtracting the practice values from the priority values. An analysis of these differences reveals insights into aspects of desirable science education of most and least dissatisfaction and thus areas that appear to be in most and least need for improvement.

A major finding of these analyses is that in general, all categories except "curriculum framework" feature positive priority-practice differences, indicating that the given aspects' importance is generally considered higher than their realization and that they are thus underrepresented in current science education. Hence, hypothesis 4b can be confirmed: For most of the aspects, their degree of realization falls short of their priority. In very few cases, categories feature negative priority-practice differences and thus overrepresentation in practice. These are related to curriculum framework as well as to scientific sub-disciplines and scientific concepts. As the latter are likely to be reflected by curriculum framework as well, this finding can be seen as an indication that science education is perceived as being too focused on the structure of the disciplines and specialized content.

Moreover, a rather large overall mean of all priority-practice differences in the total sample (M=1.2) and 65 out of 80 categories (81%) featuring differences of M>0.75 point to the stakeholders being in large parts strongly dissatisfied with current science education. Therefore, hypothesis 4a can be confirmed as well: For the majority of the aspects, considerable priority-practice differences appear. Aspects with the largest gaps between their priority and realization point towards areas that are particularly underrepresented in science education and thus seen to be in the most urgent need for improvement. Mostly, these aspects refer to skills and competences that are related to more general aims of education, student orientation, interdisciplinarity and the relation between science and society. As all of these aspects appear among those categories that feature priority values greater or equal to the mean value of all priority assessments (M=4.4), it can be concluded that highest dissatisfaction regarding current science education appears for those areas that are at the same time considered most important. Therefore, hypothesis 4c can be confirmed as well: For the highly prioritized aspects, especially large under-representation in practice can be identified. In addition to this finding, the sample shows comparably strong consensus about their degrees of dissatisfaction for most of the categories with the highest priority-practice differences on the basis of the standard deviations of the categories' individual average priority-practice differences.

In contrast, aspects that more closely match the stakeholders' expectations, while still assessed as underrepresented, mostly refer to scientific sub-disciplines and scientific concepts. A majority of these categories are amongst those aspects with the highest degree of consensus concerning the match between priority and practice.

A more differentiated analysis of the priority-practice differences is provided by considering the four sub-samples. As already foreshadowed by the science education researchers considering all given aspects on average as most important among all groups and at the same time as least realized, this group shows the highest average gap between priority and realization of the given aspects on the basis of the mean values over all priority-practice differences and thus highest dissatisfaction with the categories' representation in practice (M=1.5). In contrast, considering the given aspects with the lowest average priority and at the same time as most highly realized, the smallest gap between priority and practice appears in the group of students (M=0.8). This shows that the students consider all given aspects on average as comparably more adequately represented in the classroom than the other sub-samples and thus express lowest dissatisfaction with current science education. Hence, the students can be considered the group of stakeholders being least critical towards the current state of science education. These findings are enhanced by a consideration of the individual priority-practice differences, which feature mean values of M>0.75 for 46 out of 80 categories (58%) in the group of students. However, differences with M>0.75 appear in the group of teachers for 63 out of 80 categories (79%), in the group of education researchers for 71 out of 80 categories (89%), in the group of scientists for 68 out of 80 categories (85%), and in the adult sub-samples altogether, for 69 out of 80 categories (86%). The individual categories' standard deviations show that the students consider the given aspects with comparably more similar dissatisfaction than the other sub-sub-samples.

A more detailed analysis of the priority-practice differences in the sub-samples shows that several different accentuations in terms of areas with highest and lowest dissatisfaction can be found. With respect to statistically significant differences, 130 out of 480 possible pair comparisons (representing 27% of the cases of all pair comparisons) can be identified. Hence, hypothesis 5b, assuming that the stakeholder groups differ in their specific assessments of desirable science education can for areas identified as most deficient not fully be confirmed. Instead, in the majority of the cases, a consensus among all stakeholders regarding their extent of dissatisfaction with the imbalance between desirable science education and the current state can be identified. Yet, in more than 25% of the given aspects, the extents of dissatisfaction deviate among the sub-samples. This questions the frequently invoked consensus on scientific literacy based science education and underlines the importance of including all affected stakeholders in a reflection on desirable science education. As a result from similarities in their priority and practice assessments, the highest agreement on areas with most need for action appears between teachers and education researchers (with differences of only N=12, representing 15% of all cases of this pair comparison). Resulting from the largest number of differences in priority and practice assessments, the highest number of distinctions (N=39) can be found in a comparison between students and education researchers (49% of all cases of this pair comparison). This also reiterates tendencies from the first round.

When comparing the students with the group of adult sub-samples, 29 significant differences (36% of the cases in this group comparison) in the priority-practice differences can be found. Therefore, it can be concluded that although students and adults affected by science education generally share several similar dissatisfaction, a considerable number of deviating emphases points towards differences in their demands for improvements towards scientific literacy based science education. Yet, as these differences do not represent the majority of cases, the hypothesis of a particular gap between the views of the young and adult generation (hypothesis 5c) on desirable science education can only partly be confirmed for the priority-practice difference of the second round. However, this finding again questions the frequently invoked consensus in Germany about scientific literacy based science education and suggests a stronger integration of students' views into current discourse on improving science education.

Although all sub-samples show high demands for improvement of all aspects in the field of qualification, their priority and practice values generated most deviating priority-practice differences and thus most disagreement on the extent of dissatisfaction in the field of general and scientific competences. Therefore, this area seems to represent not only a field of desirable science education with urgent need for improvement but at the same time also seems to be in need of further discussion and discourse.

All in all, for the vast majority of the aspects, positive priority-practice differences appear, indicating that in general, their importance falls short of their realization and thus the demands posed by the participating stakeholders on scientific literacy based science education are not fulfilled in practice. According to the participating stakeholders, greatest importance should be placed on offering science education that enhances the students' abilities of critical questioning, judgement, and reflection. Moreover, it can be concluded that science education should especially put stronger emphases on enhancing skills and competences that are related to more general aims of education, address the relation between science and society, take into account more interdisciplinary approaches and should be characterized by stronger student orientation, as these areas are particularly underrepresented in science education and thus seen to be in most urgent need for improvement. A primary concern should be to find ways in practice to make a realization of these aims possible.

4.2.6   Hierarchical Cluster Analysis

On the basis of the data collected within part II of the second round question-naire, conceptions of desirable science education are identified by means of hier-archical cluster analysis. In the following parts, the response behavior in part II of the second round questionnaire is presented. Moreover, the clustering on the basis of the different category combinations as well as category frequencies within the resulting clusters are described, followed by a description of empiri-cally identified concepts of desirable science education.

4.2.6.1   Response Behavior

The results of the hierarchical cluster analysis are based on the responses on the form sheets which the participants were asked to fill out in the second part of the questionnaire. In order to identify concepts of science education that are consid-ered important and meaningful on the basis of the results from the previous round, the participants were provided with ten identical form sheets and asked to combine those aspects from the given set of 80 categories that seem especially important to them in their combination (see appendix). All in all, a total of 256 form sheets were returned by the participants in the second round (Table 23). On average, every participant filled out 1.7 form sheets. With respect to the four sub-samples, it can be seen that the highest average number of form sheets with an average of 2.7 form sheets per person is provided by the group of science educa-tion researchers, followed by the group of scientists (1.6 form sheets per person) and the group of students (1.4 form sheets per person). The lowest average num-ber of form sheets was returned by the group of teachers (1.3 form sheets per person). Relating these response profiles to the response behaviors of the first round, certain patterns become apparent: In both cases, the group of education researchers can be characterized by the highest degree of differentiation in terms of providing different category combinations of desirable science education. Lowest degrees of differentiation in both the first and second round appear in the group of students.

The majority of participants took the opportunity of using all ten form sheets for their category combinations. Instead, as shown in Table 24, 70% of the stakeholders returned two form sheets or less. Around 16% of the participants filled out one form sheet. Six participants filled out the highest possible number of form sheets (N=10).

Table 23
Response Behavior in Part II of the Second Round

Sub-Sample		Number of form sheets		Average number of form sheets	
Students		49		1.4	
Teachers	Education Students	38		1.3	
	Trainee Teachers	4	66	1.0	1.3
	Teachers	23		1.4	
	Teacher Educators	1		1.0	
Education Researchers		77		2.7	
Scientists		64		1.6	
Total		256		1.7	

Table 24
Number of Form Sheets Used by the Participants in the Second Round

Number of Form Sheets Used	Frequency	Percentage [%]	Cumulative Percentage [%]
1	24	15.6	15.6
2	85	55.2	70.8
3	25	16.2	87.0
4	8	5.2	92.2
5	1	0.6	92.9
6	2	1.3	94.2
7	1	0.6	94.8
8	1	0.6	95.5
9	1	0.6	96.1
10	6	3.9	100.0

4.2.6.2 Hierarchical Clustering on the Basis of the Category Combinations

Representing quantitative data suitable for multivariate methods, the groups of categories considered meaningful by the participants in their combination are analyzed by means of hierarchical clustering (cf. 3.2.2.3.3) to identify empirically based conceptions of desirable science education. As clusters consist of values that are characterized by a closer distance (similarity) between each other than to values in other clusters, hierarchical clustering represents a systematic distance-based method for identifying similarity structures (patterns) in a data set and thus an adequate way to structure the participants' category combinations. With the purpose of identifying concepts of desirable science education on the basis of the resulting clusters, an agglomerative clustering method is applied, starting at the bottom of the hierarchy with each object representing one individual cluster and in the course of the clustering process gradually merging pairs of clusters, which are in this way summarized into larger clusters until all objects belong to one cluster. In order to obtain in view of this aim clusters of similar sizes that are as homogeneous as possible, the Ward method is used with the squared Euclidean distance as interval scale related distance measure that is calculated between all objects. In this way, the clustering process preferably merges objects that keep the dispersion in a category group at the lowest level. Also, with advancing fusion, this method tends to balance differences in the occupation numbers of the clusters. To avoid a division of the clusters only among the patterns of the traditional scientific disciplines, categories referring to subject based contexts and scientific disciplines are not included in the hierarchical cluster analysis. As the identification of conceptions is focused on content related aspects, methodical aspects were not included in the given lists of categories for combination (cf. 3.2.2.2) and thus do not appear in the cluster analysis.

The structure of the clusters produced by the hierarchical clustering and the steps of the process are described by a dendrogram (Figure 19). The dendrogram illustrates the hierarchical partition of the data into smaller and smaller subsets as described above and is thus a compact visualization of the distance matrix. The root of the tree represents a single cluster that contains all objects of the dataset, while the leaves symbolize clusters containing single objects. Vertical lines indicate merged clusters. The position of the line on the scale illustrates the distance between objects that are clustered in a particular step. This means that the larger the distance is before two clusters are joined, the bigger the differences is between these clusters. The observed distances are rescaled to fall into a range of 0 to 25 along the top of the chart. In this way, original distances are not displayed in the dendrogram, but the ratio of the rescaled distances in the dendrogram corresponds to the ratio of the actual distances.

All original distance values and the progress of joining clusters from the first stage (every object represents a separate cluster) to the last stage (all objects are grouped into one cluster) are shown by the agglomeration table (see appendix). Each row of the table represents a clustering stage. The number of the stage is indicated in the first column. In this way, the first row describes the cluster formation at the first stage, that is, the first fusion of two individual categories into a cluster. The column titled "merged clusters" shows which objects are summarized[29]. The distance between merged objects is described in the coefficient column. As shown in the table, the categories merged first and thus those with closest proximity, are "curriculum framework" and "history of the sciences". The gradual increase of the distance values throughout all subsequent stages of the clustering results from the applied method according to which objects with the highest similarity, that is, with the shortest distance, are combined first, then objects with the second largest similarity, etc.

Table 25
Aspects not Included in the Cluster Analysis

Subject-based Contexts	Scientific Disciplines		Methods
- Science – biology - Science – chemistry - Science – physics	- Botany - Zoology - Human biology - Genetics / molecular biology - Evolutionary biol. - Neurobiology - Ecology - Microbiology - Inorg. and general chemistry	- Organic chem. - Analyt. Chem. - Biochemistry - Thermodynamics - Electrodynamics - Mechanics - Atomic and nuclear physics - Astronomy / space	- Cooperat. Learning - Learning in mixed-aged classes - Interdisciplinary learning - Inquiry-based science learning - Using new media - Learning at stations - Role play - Discussion / debate

As mini-clusters are too differentiated for interpretation in terms of conceptions, and extremely large clusters are not interpretable on a content level with sufficient precision, the selection of the number, size and type of clusters to be obtained is a crucial aspect in the interpretation of the results of a hierarchical cluster analysis. In general, the number of clusters results from defining a reasonable cutting level in the dendrogram. For this purpose, the development of the distance values over the different clustering levels has to be considered for special shifts among the fusions. It can be seen that the distance values initially increase quite moderately, while in the later stages, a stronger increase is notable. This is a typical development of distance values in the course of a clustering progress and indicates that at the lower levels of fusion, clusters with higher similarity are found, while at the later stages of merging, more dissimilar clusters are joined. Thus, while in the lower stages of merging, the mergers are rather homogeneous, in the higher stages, more heterogeneous clusters emerge. With nodes at 2087.4

---

29 The objects joined at the different clustering levels are specified in this table by sequential numbers. In this way, new numbers are assigned in the course of the process and the numbers of this column do not necessarily correspond to the categories numbers of the dendrogram.

and 2183.8, a large shift in the distance values and thus a demarcation point can be found between step 56 and 57 of clustering (cf. Figure 19). This is a first indication that reasonable clustering might end at step 56 and that the arrangement of the clusters up to this point can be used as a final result. However, as one of the final four clusters resulting from a cut at this stage contains only 7 categories, it seems more reasonable to partition at the next level (2304.4) in order to obtain more balanced clusters with the aim of identifying conceptions of desirable science education. From the partition at this level, a three-cluster solution emerges.

Table 26
Distribution of the Categories Among the Clusters of the Three-Cluster Solution

Cluster A	Cluster B	Cluster C
. Emotional personality development	. Intellectual pers. development	. Education / general personality development
. Media / current issues	. Science – interdisciplin.	. Students' interests
. Global references	. Interdisciplinarity	. Motivation and interest
. Empathy / sensibility	. Current scientific research	. Everyday life
. Perception / awareness / observation	. Scientific inquiry	. Society / public concerns
. History of the sciences	. Critical questioning	. Nature / natural phenomena
. Out-of-school learning	. Analysing / drawing concl.	. Comprehension / understanding
. Curriculum framework	. Applying knowledge / creative and abstr. thinking	. Acting reflectedly and responsibly
. Communication skills	. Content knowledge	. Judgement / opinion-forming / reflection
. Reading comprehension	. Formulating scientific questions / hypotheses	. Ethics / values
. Finding information	. Terminology	. Food / nutrition
. Social skills / teamwork	. Matter / particle concept	. Health
. Occupations / occupational fields	. Structure / function / propert.	. Medicine
. Knowledge about science-related occupations	. Chemical reactions	. Matter in everyday life
. Occupation / career	. Experimenting	. Environment
. Earth sciences	. Models	. Consequences of technol. developments
. Industrial processes	. Limits of scientific knowl.	. Safety and risks
. Cycle of matter	. Technology	. working self-dependently / structuredly / precisely
. Development / growth	. Technical devices	
	. System	
	. Interaction	
	. Energy	
	. Mathematics	
$N_{cat} = 19$	$N_{cat} = 23$	$N_{cat} = 18$
$N_{cases} = 494$	$N_{cases} = 1347$	$N_{cases} = 1335$
$N\%_{cases} = 15,6\%$	$N\%_{cases} = 42,4\%$	$N\%_{cases} = 42,0\%$

Note. N_{cat}=number of categories, N_{cases}=number of cases, $N\%_{cases}$=relative frequency regarding all cases

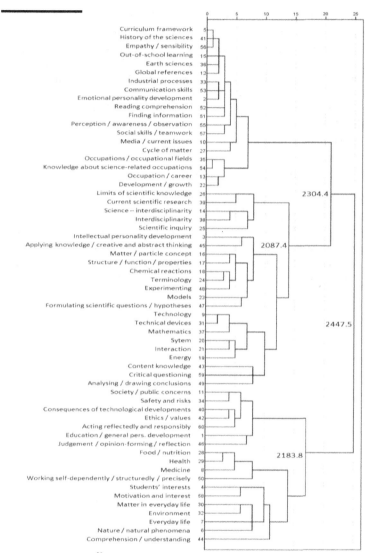

Figure 19. Dendrogram[30]

---

30 This figure can also be accessed via www.springer.com and "Theresa Schulte" within the OnlinePLUS programme.

Table 26 shows the distribution and allocation of the different categories among the three clusters, taking into account the number of categories (N_{cat}), the number of cases (N_{cases}) and the relative frequency regarding all cases ($N\%_{cases}$) in each cluster. As can be seen, 494 cases out of a total of 3176 category allocations distributed among the three clusters belong to Cluster A (15.6% of all category allocations), 1347 cases appear in Cluster B (42.4% of all category allocations), and 1335 cases in Cluster C (42.0% of all category allocations). This indicates that categories that appear in Cluster B and Cluster C are generally more often included in the combinations than those of Cluster A.

4.2.6.3 Category Frequencies in the Clusters

To illuminate the content related profiles of the clusters, Table 27 provides an overview of the distribution of the different category allocations among the three clusters. The number of cases in the different categories is taken account of in terms of their relative frequency regarding the individual clusters (f_c) as well as the relative frequency in relation to all category allocations (f_A). For the content related profiling of the clusters, especially those categories with comparably high frequencies are taken into account.

Table 27
Category Allocations in the Clusters: Total Numbers and Relative Frequencies

Category	Cluster A			Cluster B			Cluster C		
	N	f_c [%]	f_A [%]	N	f_c [%]	f_A [%]	N	f_c [%]	f_A [%]
Ed. / general pers. devel.							89	6.7	2.8
Emotional pers. devel.	27	5.5	0.9						
Intellectual pers. devel.				62	4.6	2.0			
Students' interests							97	7.3	3.1
Curriculum framework	11	2.2	0.3						
Nature / nat. phenomena							85	6.4	2.7
Everyday life							94	7.0	3.0
Medicine							67	5.0	2.1
Technology				37	2.7	1.2			
Media / current issues	39	7.9	1.2						
Society / public concerns							51	3.8	1.6
Global references	24	4.9	0.8						
Occupation / career	31	6.3	1.0						
Science – interdisc.				77	5.7	2.4			
Out-of-school learning	21	4.3	0.7						
Matter / particle concept				37	2.7	1.2			
Structure / function / prop.				51	3.8	1.6			
Chemical reactions				37	2.7	1.2			
Energy				65	4.8	2.0			
System				38	2.8	1.2			
Interaction				54	4.0	1.7			
Development / growth	26	5.3	0.8						
Models				74	5.5	2.3			
Terminology				36	2.7	1.1			
Scientific inquiry				89	6.6	2.8			
Limits of sc. knowledge				55	4.1	1.7			
Cycle of matter	36	7.3	1.1						
Food / nutrition							63	4.7	2.0
Health							64	4.8	2.0
Matter in everyday life							60	4.5	1.9
Technical devices				33	2.4	1.0			
Environment							66	4.9	2.1
Industrial processes	18	3.6	0.6						
Safety and risks							44	3.3	1.4
Occupations / occ. fields	26	5.3	0.8						
Earth sciences	21	4.3	0.7						
Mathematics				52	3.9	1.6			
Interdisciplinarity				52	3.9	1.6			
Current scientific res.				72	5.3	2.3			

Category	Cluster A			Cluster B			Cluster C		
	N	f_c [%]	f_A [%]	N	f_c [%]	f_A [%]	N	f_c [%]	f_A [%]
Consequences of technol. Developments							71	5.3	2.2
History of the sciences	16	3.2	0.5						
Ethics / values							69	5.2	2.2
Content knowledge				76	5.6	2.4			
Compr. / underst.							116	8.7	3.7
Applying knowl. / creative / abstr. think.				64	4.8	2.0			
Judgement / opinion-forming / reflection							83	6.2	2.6
Formulating scientific questions / hyp.				61	4.5	1.9			
Experimenting				47	3.5	1.5			
Analysing / drawing conclusions				94	7.0	3.0			
Working self-dep. / structuredly / prec.							75	5.6	2.4
Finding information	32	6.5	1.0						
Reading comp.	26	5.3	0.8						
Communication skills	26	5.3	0.8						
Knowledge about science-related occ.	27	5.5	0.9						
Perc./ awareness / obs.	37	7.5	1.2						
Empathy / sensibility	16	3.2	0.5						
Soc. skills / teamwork	34	6.9	1.1						
Motivation and int.							60	4.5	1.9
Critical questioning				84	6.2	2.6			
Acting reflectedly and responsibly							81	6.1	2.6
Total	494	100.0	15.6	1347	100.0	42.4	1335	100.0	42.0
Total number all cases				3176					

4.2.6.4   Cluster-Analytically Identified Conceptions of Desirable Science
Education

On the basis of the clusters identified through hierarchical clustering, three concepts of desirable science education are identified. In the following the concepts
are described.

**Concept A: Awareness of the sciences in current, social, globally relevant
and occupational contexts relevant in both educational and out-
of-school settings**

Concept A refers to an engagement with the sciences within the frame of current,
social, globally relevant, occupational and both educational and out-of-school
contexts, enhancing emotional personality development and basic skills. The
impressions a person gets through engaging with topics and associated science
related questions from his or her environment influence both the person's sensibility and his or her attitudes towards the sciences. Dealing with scientific issues
or phenomena in out-of-school or social and public contexts respectively also
facilitates conscious experiences of scientific phenomena, scientifically precise
observation and cognitive ability. Moreover, basic and professionally relevant
skills such as finding, interpreting and communicating information can be enhanced in this way. Suggestions for this kind of engagement and education are
amongst others provided e.g. by current issues or media coverage. Dealing with
the history of the sciences especially reveals how findings and methods of the
sciences enable, enhance and bring forward research in science. This shows
moreover how historical science related developments are linked to applications
in industry and technology, how these applications changed the world and how
they influence our professional and everyday lives.
The following categories constitute Concept A:

Situations, contexts, motives:
Emotional personality development, media / current issues, global references,
occupation / career, out-of-school learning, curriculum framework

(Basic) concepts, themes and perspectives:
History of the sciences, occupations / occupational fields, industrial processes,
cycle of matter, earth sciences, development / growth

Qualifications:
Empathy / sensibility, perception / awareness / observation, social skills / team-
work, knowledge about science-related occupations, communication skills, finding information, reading comprehension

Concept B: Intellectual education in interdisciplinary scientific contexts

Concept B refers to an engagement with science, its terminology, methods, basic
concepts, interdisciplinary relations, findings, and its perspectives, which enhance individual intellectual personality development. Dealing with science
serves in this course not only the acquisition of science-related basic knowledge
but also helps to understand fundamental discoveries and the process of gaining
knowledge in the sciences. Moreover, dealing with questions and topics of science helps to comprehend and follow (empirical and experimental) scientific
research methods, facilitates analytical skills and fosters the ability to take differentiated perspectives. In addition, an engagement with current scientific research
reveals not only how findings and methods of the sciences enable, enhance and
support both scientific research and its applications, but also how scientific research is interconnected on interdisciplinary levels.
The following categories constitute Concept B:

Situations, contexts, motives:
Intellectual personality development, science - interdisciplinarity, technology

(Basic) concepts, themes and perspectives:
Interdisciplinarity, scientific inquiry, current scientific research, limits of scientific knowledge, terminology, matter / particle concept, structure / function /
properties, chemical reactions, models, technical devices, system, interaction,
energy, mathematics

Qualifications:
Applying knowledge / creative and abstract thinking, formulating scientific questions / hypotheses, factual knowledge, critical questioning, analyzing / drawing
conclusions, experimenting

**Concept C: General science-related education and facilitation of interest in
contexts of nature, everyday life and living environment**

Concept C refers to science related engagement with everyday life and living
environment issues that takes up and promotes students' interests, enhancing

generally their personality development and education. In this context, aspects such as opinion-forming and acting reflectedly and responsibly play a particularly important role. Dealing with topics from the natural and technological environment shows how scientific research, scientific applications and scientific phenomena influence both public and personal life. Another important aspect of this concept is engaging with different values and perspectives as well as reflecting on both personal and public discourse, decisions and actions. Moreover, this concept refers to facilitating the motivation for dealing with scientific inquiry beyond school, including aspects such as realizing and shaping one's own interests. Engaging with scientific issues and phenomena within the frame of social and public fields such as technological developments, their consequences and issues about safety and risks enhances in particular the students' abilities to judge and both critically reflect and rationally account for their own actions. The following categories constitute Concept C:

Situations, contexts, motives:
Society / public concerns, students' interests, education / general personality development, nature / natural phenomena, everyday life, medicine

(Basic) concepts, themes and perspectives:
Safety and risks, consequences of technological developments, ethics / values, food / nutrition, health, matter in everyday life, environment

Qualifications:
Acting reflectedly and responsibly, judgement / opinion-forming / reflection, motivation and interest, comprehension / understanding, working self-dependently / structuredly / precisely.

4.2.6.5   Discussion

With a total of 256 form sheets returned by the participants, the data gathered in part II of the second round provides a satisfying basis for the applied hierarchical cluster analysis. The resulting clusters contain 19 (Concept A), 23 (Concept B) and 18 (Concept C) categories respectively and are thus adequately balanced with regard to the distribution of the different categories. With respect to the aim of identifying conceptions of desirable science education, the chosen method of hierarchical clustering proved to be applicable to determine structures in the participants' responses. On the basis of the three empirically established clusters, three concepts of desirable science education are identified. Therefore, by means of the Delphi method, a concretion and synthesis of the participants' views of desirable science education was reached within part II of the second round. The identified concepts relate to an "awareness of the sciences in current, social, globally relevant, occupational contexts relevant in both educational and out-of-school contexts" (Concept A), "intellectual education in interdisciplinary scientific contexts" (Concept B), and "general science-related education and facilitation of interest in contexts of nature, everyday life and living environment" (Concept C). As the concepts are the result of a clustering process based on proximities and distance similarities, it is important to note that the three concepts are not to be understood as mutually exclusive concepts of desirable science education but rather as concepts with different emphases enhancing each other.

In what way the three concepts are assessed in terms of their priority and realization in practice, both in science education in general and depending on different levels of education, and in what ways the different sub-samples' assessments differ from each other, is shown by an analysis of the results obtained within the third round

4.3 Results of the Third Round

In the following sections, I will present the results of the third and final round of the Berlin Curricular Delphi Study in Science. The results of the third round provide insights about how the three identified concepts of desirable science education are assessed in terms of their priority and realization in practice. Also, the results indicate whether the concepts receive at different educational levels within the frame of general education similar or different priorities and how far these concepts are sufficiently realized at the different levels of education.

First, I will describe the sample and the response rate. Again, according to the two-part structure of the third round questionnaire (cf. 3.2.3.2), the description of the results is divided into two parts. The first part (4.3.2) includes an analysis of the priority and practice assessments as well as priority-practice differences with respect to science education in general; the second part (4.3.3) addresses the assessments according to different levels of education within the frame of general education.

4.3.1 Sample and Response Rate

In accordance with the Delphi method, which is based on a fixed group of participants, only the 154 stakeholders who participated in the second as well as the first round of the Berlin Curricular Delphi Study in Science were asked to fill in the questionnaire of the third round. The data collection, again in paper-and-pencil as well as electronic form, took place between July and November 2012.

The sample structure of the third round and a comparison of the response rates between the second and third round as well as the first and third round are shown in Table 28. The drop-out over the three rounds is visualized in more detail in Figure 20. With regard to drop-out between the second and third round, it can be seen that out of the 154 participants from the second round, a total of 109 participants took part in the third round. Hence, 109 participants took part in all three rounds. This corresponds to 71% of the participants from the second round and 56% of the participants from the first round.

With respect to the sub-samples, it can be seen that the group of science education researchers features the highest response rate (83%) in the third round. With slightly higher drop-out, the participation rate between the second and third round in the group of students is 76% and in the group of scientists 71%. The highest drop-out between the second and third round appears in the group of teachers with a response rate of 60%. When comparing the drop-out progression of the sub-samples over the three rounds (Figure 20), it becomes apparent that the teachers and scientists feature a relatively strong decline in participation numbers, while students and science education researchers seem to be the more consistent participants.

A detailed overview of the sample structure in the third round of the Berlin Curricular Delphi Study in Science is provided in Table 29. With 26 participants, the students make up 24% of the total sample. The group of teachers constitutes the largest group with 30 participants (28% of the total sample). The group of science education researchers represents the smallest part of the sample with 24 participants and 22% of the total sample. The group of scientists is the second largest part of the sample with 29 participants and 27% of all participants in the third round.

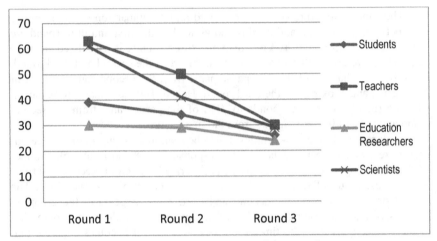

Figure 20. Drop-out of participants in the sub-samples over all three rounds

Table 28
Overall Sample Structure of the Berlin Curricular Delphi Study in Science and Response Rates

Sub-Sample		Number of Participants			Participation rate between rounds	Participation rate between rounds
		Round 1	Round 2	Round 3	2 and 3	1 and 3
Students		39	34	26	76%	67%
Teachers	Education Students	32	29	10		
	Trainee Teachers	5	4	4		
	Teachers	18 63	16 50	16 30	60%	48%
	Teacher Educators	8	1	0		
Education Researchers		30	29	24	83%	80%
Scientists		61	41	29	71%	48%
Total		193	154	109	71%	56%

Table 29
Detailed Structure of the Sample in the Third Round

Sub-sample	Sub-group	Subject	Distribution[31]	Total Number	Percentage
Students	Basic Science Courses	Biology	15		
		Chemistry	13		
		Physics	15	26	23.8%
	Advanced Science Courses	Biology	3		
		Chemistry	10		
		Physics	2		
Science Teachers	Science Education Students at University	Biology	4		
		Chemistry	8	10	
		Physics	2		
		Science (elem. level)	1		
	Trainee Science Teachers	Biology	0		
		Chemistry	4	4	
		Physics	0		
		Science (elem. level)	1		30	27.5%
	Science Teachers	Biology	4		
		Chemistry	9	16	
		Physics	5		
		Science (elem. level)	1		
	Trainee Science Teacher Educators	Biology	0		
		Chemistry	0	0	
		Physics	0		
		Science (elem. level)	0		
Science Education Researchers		Biology	4		
		Chemistry	12		
		Physics	6	24	22.0%
		Science (elem. level)	0		
		Not specified	2		
Scientists		Biology	5		
		Chemistry	12		
		Physics	8	29	26.6%
		Others	5		
		Not specified	2		
Total				109	–

---

31 As multiple answers per participant with regard to these characteristics were possible, the numbers in this column do not represent summands of the total numbers of participants in the corresponding sample groups.

4.3.2   General Assessment

A consideration of the three concepts in terms of their priority and practice assessments as well as priority-practice differences with regard to science education in general and a comparison between the different sub-samples reveals a more condensed picture of the participating stakeholders' expectations towards science education in Germany, their views of the reality of science education, and areas of highest dissatisfaction.

A suitable representation of the stakeholders' opinions can be derived from on an analysis of the mean values as well as a consideration of standard deviations as indicators of the degree of consensus among the participants concerning their assessments. Again, the analyses are carried out for both the total sample and the four different sub-samples (students, teachers, educations researchers and scientists).

4.3.2.1   Priority Assessments

4.3.2.1.1   Total Sample

The following part addresses the assessments by the total sample with respect to the concepts' priority in science education in general, for which the mean values are displayed in Table 30. Since the concepts are, as in the second round, assessed on a scale from 1 ("very low priority") to 6 ("very high priority"), values over the theoretical mean value of 3.5 can be interpreted as the corresponding concept being considered as at least fairly important. All mean values range above 4, which shows that on average, all sub-samples assess the concepts above the theoretical mean value of 3.5 and with at least "rather high priority".

The highest mean value (M=5.0; "high priority") appears for Concept C, which refers to general science-related education and facilitation of interest in contexts of nature, everyday life and living environment. A mean value of M=4.7 is found for Concept A, which relates to an awareness of the sciences in current, social, globally relevant and occupational contexts relevant in both educational and out-of-school settings. Concept B, which refers to intellectual education in interdisciplinary scientific contexts, features a mean value of M=4.3. The significance values of the Wilcoxon signed-rank test with respect to the pair comparisons of the concepts (Table 30) show that the assessments of the concepts differ in all three pair comparisons from each other in a statistically significant way.

With standard deviations in the priority assessments only in some cases slightly surpassing a value of SD=1.0, it can be seen that the stakeholders broadly agree on the importance of the empirically identified concepts of desirable science education. The highest degree of agreement appears for the most highly prioritized concept (Concept C).

Table 30
Mean Values and Significant Differences (Wilcoxon Signed-Rank Test) of the General Priority Assessments by the Total Sample

		Mean Values		Significant Differences	
Concept A: Awareness of the sciences in current, social, globally relevant and occupational contexts relevant in both educational and out-of-school settings	Concept B: Intellectual education in interdisciplinary scientific contexts	Concept C: General science-related education and facilitation of interest in contexts of nature, everyday life and living environment	A/B	A/C	B/C
4.7	4.3	5.0	*	*	*

Note. *=significant (p<0.05)

4.3.2.1.2 Sub-Samples

Insights into differences in assessment by the sub-samples can be gained through a consideration of the average assessments of all three concepts in the different sub-samples (Figure 21). The highest priorities are given by the group of science education researchers (M=5.1), followed by the teachers (M=4.8) and scientists (M=4.6). The lowest priorities are provided by the students (M=4.2). In this way, the four sub-samples assess all three categories together with at least "rather high priority". This sequence corresponds to the sub-samples' order in the priority assessments of the second round. When comparing the students and the group of adult sub-samples, a tendency towards higher overall priorities can be found in the group of adult sub-samples (M=4.8).

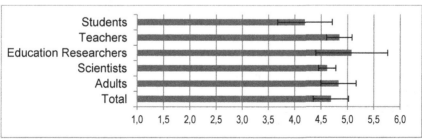

Figure 21. Overall mean values of the concepts' priority assessments in the sub-samples

The extent of differentiation among the priorities of the individual concepts is shown through the standard deviations as indicated by error bars in Figure 21. The lowest degree of differentiation, thus most similar priorities for the concepts, can be found in the group of scientists (SD=0.16). The highest standard deviation (SD=0.68) can be found in the group of education researchers, which implies that participants in this group differentiate to a higher degree between their priority assessments than the other groups. Yet, as shown by the pair comparisons of the concepts, it can be seen that in most sub-samples, the concepts are not prioritized on significantly different levels.

Further, more detailed comparisons of the priority assessments by the different sub-samples show that although in general, the different sub-samples assess the concepts with similar priorities, the sub-samples differ in a number of pair comparisons from each other (Table 31). In total, 7 statistically significant differences out of 18 possible pair comparisons among the four sub-samples can be found (38% of all priority related pair comparisons). The most differences appear for Concept A (N=4). A comparison of the students and the group of adult sub-samples reveals significant differences for 2 out of 3 concepts. This indicates different emphases between the students and adult sub-samples.

A comparison of the mean values with normalized values reveals very small deviations between absolute values of $\Delta T=0.00$ and $\Delta T=0.02$ (see appendix). This shows that the differences between the sizes of the sub-samples do not cause a distortion of the mean values.

Table 31
Significant Differences (Mann-Whitney-U-Test) and Mean Values of the General Priority Assessments by the Sub-Samples

Con-cepts	Significant Differences							Mean Values					
	S/T	S/E	S/Sc	T/E	T/Sc	E/Sc	S/A	S	T	E	Sc	A	Total
Concept A	*	*	*			*	*	3.8	5.0	5.3	4.7	5.0	4.7
Concept B	*						*	4.0	4.6	4.3	4.4	4.4	4.3
Concept C		*				*		4.8	5.0	5.6	4.7	5.1	5.0
All Con-cepts	2	2	1	0	0	2	2	4.2	4.8	5.1	4.6	4.8	4.9
			7										

Note. S=students, T=teachers, E=education researchers, Sc=scientists, A=adults (teachers, education researchers, and scientists), *=significant (p<0.05)

4.3.2.2   Practice Assessments

4.3.2.2.1   Total Sample

In addition to the priority assessments, the following section focuses on the total samples' assessments of the concepts' realization in science education in general. The mean values of the assessments are displayed in Table 32. Since the concepts are, as in the second round, assessed on a scale from 1 ("to a very low extent") to 6 ("to a very high extent"), values over the theoretical mean value of 3.5 can be interpreted as the concepts being considered at least fairly present in science education in general. In contrast to the priority assessments and similar to the practice assessments of the second round, it can be noticed that none of the three concepts is assessed with practice values above 3.5.

Table 32
Mean Values and Significant Differences Values (Wilcoxon Signed-Rank Test) of the General Practice Assessments by the Total Sample

Mean Values			Sigificant Differences
Concept A: Awareness of the sciences in current, social, globally relevant and occupational contexts relevant in both educational and out-of-school settings	Concept B: Intellectual education in interdisciplinary scientific con-texts	Concept C: General science-related education and facilitation of interest in contexts of nature, everyday life and living environment	A/B A/C B/C
2.9	3.1	3.3	*

Note. *=significant (p<0.05)

All three concepts are considered by the participants as not very present in science education in general. The lowest realization is assigned to an "awareness of the sciences in current, social, globally relevant and occupational contexts relevant in both educational and out-of-school settings" (Concept A; M=2.9), while the highest, yet still rather low realization is ascribed to Concept C (M=3.3), which refers to "general science-related education and facilitation of interest in contexts of nature, everyday life and living environment". The realization of Concept B ("intellectual education in interdisciplinary scientific contexts") is assessed with a mean value of 3.1. In this way, all concepts are assessed as realized to a "rather low extent".

The Wilcoxon signed-rank test with respect to pair comparisons of the concepts (Table 32) reveals that, in contrast to the priority assessments, only Concepts A and C differ significantly in their assessment in terms of realization. With standard deviations between SD=0.95 and SD=1.18, it becomes apparent that the stakeholders mostly agree on the extent of realization of the empirically identified concepts of desirable science education. The lowest standard deviation among the three concepts appears for Concept A (SD=0.95), which is the concept assessed as least present in the science classroom.

4.3.2.2.2 Sub-samples

A first approach towards a more differentiated view on the practice assessments by the sub-samples can be gained through considering the average practice assessments of all three concepts in the different sub-samples (Figure 22). The highest realization of the concepts is perceived by the group of teachers (M=3.3), followed by the education researchers (M=3.2) and students (M=3.1). The group of scientists assesses the concepts on average considerably lower (M=2.7). A comparison of the students and the group of adult sub-samples (M=3.1) shows that these two groups' opinions on the realization of the concepts mostly match.

The extent of differentiation between the assessments of the concepts is shown by the standard deviations, which are indicated by error bars in Figure 22. The lowest degree of differentiation in the practice assessments among the sub-samples and thus most similar practice values for the three concepts can be found in the group of scientists (SD=0.14). In the same way as in the priority assessments of the concepts, the highest standard deviation (SD=0.29) appears in the group of education researchers, which implies that participants in this group again differentiate to a higher degree between their practice assessments than the other groups. However, as indicated by the pair comparisons of the concepts, the majority of pair comparisons in the sub-samples do not feature significant differences.

A more detailed view on the comparisons of the practice assessments of the different sub-samples confirms the similarity of the participants' opinions regarding the realization of the concepts in practice, since in contrast to the priority assessments, none of the 18 possible pair comparisons of the concepts among the four sub-samples features significant differences (Table 33). This indicates that the sub-samples mostly agree with each other about the presence of the concepts in science education in general.

A comparison of the total samples' practice values to normalized values shows very small deviations with differences between ΔT=0.00 and ΔT=0.01

(see appendix). Again, this shows that despite uneven sub-sample distribution, the results are adequately represented by the mean values.

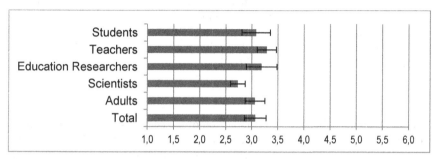

Figure 22. Overall mean values of the concepts' practice assessments in the sub-samples

Table 33
Significant Differences (Mann-Whitney-U-Test) and Mean Values of the General Practice Assessments by the Sub-Samples

Con-	Significant Differences							Mean Values					
cepts	S/ T	S/ E	S/ Sc	T/ E	T/ Sc	E/ Sc	S/ A	S	T	E	Sc	A	Total
Concept A								2.8	3.1	2.9	2.7	2.9	2.9
Concept B								3.1	3.3	3.3	2.6	3.1	3.1
Concept C								3.3	3.5	3.4	2.9	3.3	3.3
All Concepts	0	0	0	0	0	0	0	3.1	3.3	3.2	2.7	3.1	3.1

Note. S=students, T=teachers, E=education researchers, Sc=scientists, A=adults (teachers, education researchers, and scientists), *=significant (p<0.05)

4.3.2.3   Priority-Practice Differences

After analyzing the opinions of different stakeholders on the importance and degree of realization of the three concepts of desirable science education, the following section addresses a comparison of these two perspectives to gain more specific insights about areas of science education that are in need for improvement. The comparisons of the priority and practice assessments are carried out on the basis of priority-practice differences. These differences are determined by subtracting the practice value from the priority value obtained from the assessments on a six-tier rating scale. In case of a positive difference value, the corresponding concept can be considered less realized than demanded by its priority assessment and thus underrepresented in practice. In contrast, a negative difference value indicates an overrepresentation of the concept. Large positive difference values between priority and practice assessments point to great dissatisfaction between current realization and priority, while small values indicate a more balanced relation between wish and reality.

4.3.2.3.1 Total Sample

The following part focuses on a comparison of the concepts' priority and realization in science education in general assessed by the total sample. The comparison is carried out by considering the average priority-practice differences of the three concepts, which were assessed on scales from 1 ("very low priority" / "to a very low extent") to 6 ("very high priority" / "to a very high extent"). As shown in Table 34, all three concepts feature positive priority-practice differences. This makes clear that all practice assessments of the concepts are considerably smaller than the priority assessments, indicating that the participants consider all three concepts as realized to a lower extent than their priority requires.

Table 34
Mean Values and Significant Differences (Wilcoxon Signed-Rank Test) in the Priority-Practice Differences in the General Assessments of the Total Sample

Mean Values			Significant Differences		
Concept A: Awareness of the sciences in current, social, globally relevant and occupational contexts relevant in both educational and out-of-school settings	Concept B: Intellectual education in interdisciplinary scientific contexts	Concept C: General science-related education and facilitation of interest in contexts of nature, everyday life and living environment	A/B	A/C	B/C
1.9	1.3	1.7	*		*

Note. *=significant (p<0.05)

The largest difference (M=1.9) occurs for an "awareness of the sciences in current, social, globally relevant and occupational contexts relevant in both educational and out-of-school settings" (Concept A), followed by "general science-related education and facilitation of interest in contexts of nature, everyday life and living environment" (Concept C; M=1.7). A smaller gap between priority and practice (M=1.3) appears for "intellectual education in interdisciplinary scientific contexts" (Concept B). The results of the Wilcoxon signed-rank test (Table 34) reveal significant differences between Concept B and the other two concepts. This shows that Concept A and Concept C are considered as more similarly realized than Concept B.

4.3.2.3.2. Sub-Samples

In addition to the analysis of the priority-practice differences in the total sample presented above, the following section provides a more differentiated view about the discrepancies between the concepts' importance and their realization in practice through a consideration of the priority-practice differences in the sub-samples.

The mean values of the priority-practice differences of all three concepts in the four sub-samples (Figure 23) show in which sub-samples the highest overall dissatisfaction with respect to the concepts' priority and practice relation can be found. It can be seen that all four sub-samples feature positive difference values between M=1.1 and M=1.9, thus showing general dissatisfaction with the current realization of the concepts compared with their priority. The highest discrepancy between priority and realization of the concepts can be found in the groups of both science education researchers and scientists (M=1.9). This shows that these two groups express the highest dissatisfaction with the concepts' representation in science education in general. The education researchers show particularly strong dissatisfaction with the realization of Concept A. In contrast and in the same way as in the second round, the smallest gap between priority and practice appears in the group of students (M=1.1). This indicates that the students consider the concepts as more adequately represented in practice than other adult sub-samples and thus express lowest dissatisfaction with current science education.

The standard deviations, as indicated by error bars in Figure 23, illustrate the scope of the concepts' priority-practice differences in the different sub-samples. It can be seen that the highest standard deviation appears in the group of education researchers (SD=0.71). This indicates that the degrees of dissatisfaction with the priority-practice relation of the individual concepts differ most from each other in this group. The lowest deviations in the priority-practice differences can be found in the group of scientists (SD=0.13), which shows that the degrees of dissatisfaction with the realization of the individual concepts are more similar in the group of scientists than in the other sub-samples.

Yet, only 3 out of 12 possible pair comparisons of the concepts show significant differences. This means that the concepts feature mostly similar discrepancies within the sub-samples. A certain similarity between the sub-samples regarding the dissatisfaction with the concepts' realization is shown by the results from the pair comparisons of the sub-samples, which reveal only 4 out of 18 possible pair comparisons (22%) with significant differences (Table 35). The students and adult sub-samples only differ significantly in the priority-practice assessments of Concept A.

Table 35
Significant Differences (Mann-Whitney-U-Test) and Mean Values of the Priority-Practice Differences in the General Assessments by the Sub-Samples

Con-cepts	Significant Differences							Mean Values					
	S/ T	S/ E	S/ Sc	T/ E	T/ Sc	E/ Sc	S/ A	S	T	E	Sc	A	Total
Concept A	*	*	*				*	1.0	1.9	2.5	2.0	2.1	1.9
Concept B			*					0.8	1.3	1.1	1.8	1.4	1.3
Concept C								1.5	1.5	2.1	1.8	1.8	1.7
All Con-cepts	1	1	2	0	0	0	1	1.1	1.6	1.9	1.9	1.8	1.7
			4										

Note. S=students, T=teachers, E=education researchers, Sc=scientists, A=adults (teachers, education researchers, and scientists), *=significant (p<0.05)

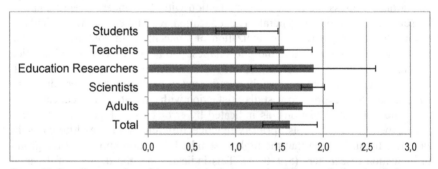

Figure 23. Overall mean values of the concepts' priority-practice differences in the sub-samples

A comparison of the total samples' priority-practice differences to normalized values shows very small deviations with values between $\Delta T=0.00$ and $\Delta T=0.02$ (see appendix).

4.3.3   Assessment According to Different Educational Levels

In addition to the results from the general assessment of the three concepts with regard to their priority, their realization in practice and priority-practice differences, this chapter addresses how the concepts are assessed according to the following levels of education:

- pre-school
- elementary level
- lower secondary education
- higher secondary education

The analyses focus on the total sample and are structured again according to priority, practice, and priority-practice differences.

4.3.3.1   Priority Assessments

In the following section, the priority assessments of the three concepts by the total sample with regard to different educational levels are presented. Table 36 shows the priorities stakeholders attribute to Concept A, Concept B and Concept C at pre-school, elementary, lower secondary and higher secondary level. In general, as indicated by the overall priority mean values of the three concepts for each of the specific educational levels (Figure 24), it can be seen that the given priorities increase with the level of education, which means that the concepts are assessed the more important the higher the educational level is.

Starting with priority mean values between M=3.3 and M=4.1 at pre-school level, ranging thus between "rather low priority" and "rather high priority", the priorities for higher secondary education reach almost values of M=5, which represents high priority. As with the general priority assessments (cf. 4.3.2.1), Concept B is given lowest importance at all levels of education. The highest priority mean values for science education at pre-school (M=4.1) and elementary level (M=4.6) appear for Concept C. For lower secondary education, Concept C and Concept A are assessed at equally high levels (M=4.8). For higher secondary education, Concept A (M=5.1) slightly surpasses Concept C (M=4.9). It can be noticed that the mean values of Concept A for pre-school and Concept B for pre-school and elementary level range below the theoretical mean value of M=3.5 and are therefore considered as not so important for these educational levels.

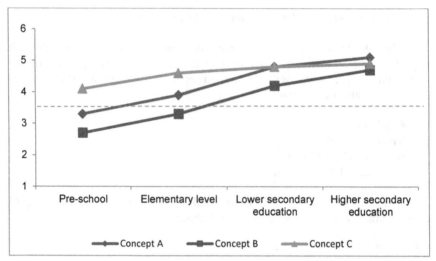

Figure 24. Mean values of the priority assessments by the total sample regarding different levels of education

Except for three pair comparisons (Concepts A/C at lower and higher secondary education and Concepts B/C at higher secondary education), the concepts differ from each other in their priorities in a statistically significant way at all educational levels. However, it can be seen that the concepts' priorities converge with higher levels of education. This means that, at higher levels of education, the concepts' are assessed as more similarly important. At the lower educational levels, the standard deviations of the priority assessments are smaller, while the highest degree of consensus for the importance of the given concepts can be found at lower secondary level.

Table 36
Mean Values and Significant Differences (Wilcoxon Signed-Rank Test) of the Priority Assessments by the Total Sample Regarding Different Educational Levels

Level of Education	Concept A: Awareness of the sciences in current, social, globally relevant and occupational contexts relevant in both educational and out-of-school settings	Concept B: Intellectual education in interdisciplinary scientific contexts	Concept C: General science-related education and facilitation of interest in contexts of nature, everyday life and living environment	Average of all three concepts	Significant Differences		
	M	M	M	M	A/ B	A/ C	B/ C
Pre-school	3.3	2.7	4.1	3.4	*	*	*
Elementary level	3.9	3.3	4.6	3.9	*	*	*
Lower secondary education	4.8	4.2	4.8	4.6	*		*
Higher secondary education	5.1	4.7	4.9	4.9	*		

Note. M=mean, *=significant (p<0.05)

4.3.3.2 Practice Assessments

In the following section, the practice assessments of the three concepts by the total sample according to different educational levels are shown. Figure 25 visualizes the perceived degrees of realization of the three concepts at pre-school, elementary, lower secondary and higher secondary level. As with the priority assessments, the extent of realization of all three concepts increases – yet not as strongly as the priority assessments – with the level of education. This progression is also illustrated by the overall practice mean values of the three concepts for each of the educational levels (Table 37). Furthermore, again similar to the tendencies in the priority assessments, the practice assessments converge with the level of education towards all three concepts being seen as realized to almost the same extent at higher secondary education.

Practice values start between levels of M=1.9 and M=2.6 at pre-school, which point to "low" or "rather low" degrees of realization. The highest practice assessments do not surpass values of M=3.5, which indicates that none of the concepts is seen at any level more than fairly realized. Concept B is assessed with lowest realization at all levels of education. Concept C shows the highest practice mean values for science education at pre-school (M=2.6), elementary level (3.1) and lower secondary level (3.5). For higher secondary education, Concept C and Concept A are assessed equally high (3.5). As shown by the results of the significance test (Table 37), Concept A and Concept B are considered present to very similar degrees at all levels of science education, but to a lower extent than Concept C. The lowest standard deviation and thus highest degree of agreement regarding the realization of the given concepts can be found at lower secondary level.

Table 37.
Mean Values and Significant Differences (Wilcoxon Signed-Rank Test) of the Practice Assessments by the Total Sample Regarding Different Educational Levels

Level of Education	Concept A: Awareness of the sciences in current, social, globally relevant and occupational contexts relevant in both educational and out-of-school settings	Concept B: Intellectual education in interdisciplinary scientific contexts	Concept C: General science-related education and facilitation of interest in contexts of nature, everyday life and living environment	Average of all three concepts	Significant Differences		
	M	M	M	M	A/ B	A/ C	B/ C
Pre-school	2.0	1.9	2.6	2.2		*	*
Elementary level	2.6	2.4	3.1	2.7		*	*
Lower secondary education	3.2	3.1	3.5	3.3		*	*
Higher secondary education	3.5	3.4	3.5	3.5			

Note. M=mean, *=significant (p<0.05)

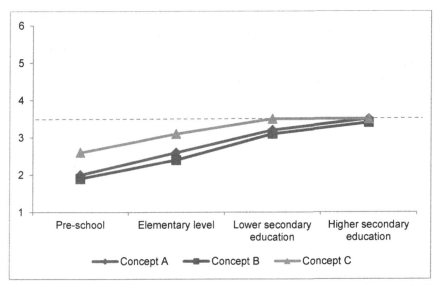

Figure 25. Mean values of the practice assessments by the total sample regarding different levels of education

4.3.3.3 Priority-Practice Differences

The following part presents a comparison between the total sample's priority and practice assessments according to different educational levels. For this purpose, the priority-practice differences of the three concepts at pre-school, elementary, lower secondary and higher secondary level are taken into account (Table 38). While both priority and practice values increase with higher levels of education, the practice values are at all educational levels considerably lower than the priority values. With mean values above $M=1.0$ for all three concepts at all levels of education, the priority-practice differences show that the overall realization of the three concepts falls short of their assigned importance in all cases. This indicates dissatisfaction with the realization of the concepts at all levels of education. As the priority and practice assessments of the three concepts both converge from pre-school to higher secondary education (cf. 4.3.3.1 and 4.3.3.2), the differences also converge (Figure 26). It is noticeable that the priority-practice differences slightly increase with the level of education. This shows that the realization of the concepts is considered more deficient with increasing levels of education. Concept C shows the highest priority-practice differences at pre-school

(M=1.5) and elementary level (M=1.4), whereas Concept A features slightly higher differences at lower (M=1.6) and higher secondary level (M=1.6). Yet, Concept C and Concept A show mostly similar differences at the different levels of education. Concept B shows the smallest priority-practice differences of the three concepts at all educational levels.

Table 38
Mean Values and Significant Differences (Wilcoxon Signed-Rank Test) between the Priority-Practice Differences in the Assessments Regarding Different Levels of Educational by the Total Sample

Level of Education	Concept A: Awareness of the sciences in current, social, globally relevant and occupational contexts relevant in both educational and out-of-school settings	Concept B: Intellectual education in interdisciplinary scientific contexts	Concept C: General science-related education and facilitation of interest in contexts of nature, everyday life and living environment	Average of all three concepts	Significant Differences		
	M	M	M	M	A/B	A/C	B/C
Pre-school	1.2	0.7	1.5	1.1	*	*	*
Elementary level	1.3	0.8	1.4	1.2	*		*
Lower secondary education	1.6	1.0	1.4	1.3	*		*
Higher secondary education	1.6	1.3	1.4	1.4	*		

Note. M=mean, *=significant (p<0.05)

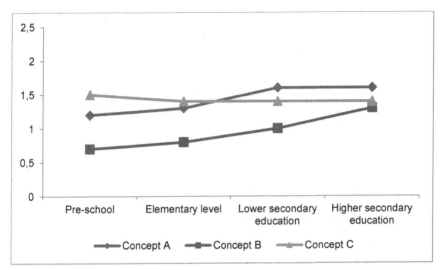

Figure 26. Mean Values of the Priority-Practice Differences in the Assessments by the Total Sample
Regarding Different Levels of Education

4.3.4 Discussion

With a return for participation of 71% between the second and third round, the
third and final round of this study includes a total of 109 stakeholders who returned for participation from the second round. This represents a satisfying sample size and drop-out rate for the final round of a long-term study such as the
Berlin Curricular Delphi Study in Science.

In the following part, the results of the third round are discussed with respect to their significance for science education in Germany. The discussion is
divided into two parts according to the assessment of the concepts regarding
science education in general and different levels of education.

General Assessment

The analyses of the general assessment of the three concepts that were established from category combinations by the stakeholders in the second round show
that the participants assess the empirically identified concepts of desirable science education with rather high or high priority. This implies that all three concepts are seen as important for science education and shows that by applying the
Delphi method, a synopsis of the stakeholders' opinions regarding desirable

science education can be derived. Furthermore, this result also verifies the solidity of the results from the previous round and further proves the applicability of the Delphi method within this study.

The highest priority is assigned to the general and everyday life related concept of science education (Concept C), followed by the concept related to an awareness of the sciences in different contexts (Concept A) and the concept referring to intellectual education (Concept B). The assessment of Concept C features the lowest standard deviations, highest degree of agreement, which means that the participants largely agree on considering this concept as most important for scientific literacy based science education. As aspects connected to the structure of the science disciplines, specialized fields, and traditional approaches of single-subject-orientation are mostly included in Concept B, hypothesis 2b, stating that these elements have lowest priority, can be confirmed. As only some of the aspects assumed in hypothesis 2a to be most highly prioritized appear in Concept C, this hypothesis can partly be confirmed.

In contrast to the priority assessments, all three concepts are considered by the participants as not very present in science education. The evaluation of the practice assessments shows that the everyday life related concept (Concept C) is considered most realized, yet only fairly present, in practice. As this concept includes aspects related to students' living environment, ethical references, and overarching aims of general education, hypothesis 3b, assuming that these aspects are perceived as realized in current science education to the lowest extents, cannot be confirmed. Aspects related to the structure of the science disciplines, specialized fields, and traditional approaches of single-subject-orientation received the highest perceived extent of realization in the second round. These aspects are mostly included in Concept B, which is not considered to be most present in practice. Therefore, hypothesis 3a cannot be confirmed. The analysis of the standard deviations shows that the participants agree most on assessing Concept A as the least realized concept in science education.

With comparably high priorities and rather low practice assessments of the three concepts, the priority-practice differences indicate large gaps between the concepts' priority and extent of realization in practice and thus show that all three concepts fall short of their given importance for science education in general. Hence, hypothesis 4a, assuming considerable priority-practice differences, and hypothesis 4b, proposing that reality is not living up to the stakeholders' expectations, can be confirmed in the third round as well. The concept considered as most deficient in its priority-practice relation is the concept related to awareness of the sciences in different contexts (Concept A). Therefore, there seems to be a most urgent need for change in the context of this concept. This finding confirms the hypothesis that for high priorities, especially large under-representations in practice can be identified (hypothesis 4c). The lowest priority-practice difference appears for the concept related to intellectual education (Concept B). The results of the pair comparisons between the concepts show that Concept B is seen by the participants as less deficient than Concept A and Concept C. Concept B is perceived as more adequately realized in current science education than the other concepts. This corresponds to findings from the second round, in which many of the aspects included in Concept B were assessed similarly.

While Concept B places an emphasis on content knowledge and intellectual education in interdisciplinary scientific contexts, the other two concepts refer to a large variety of contexts, topics and skills with current, social, globally relevant and occupational references (Concept A) as well as aspects in contexts of nature, everyday life and living environment (Concept C). Hence, it can be said that from the perspective of the German stakeholders involved in this study, science education with references to a variety of contexts is seen as more important than traditional scientific contents. This implication can also be related to claims in Europe for more practice-related and context-based science education with emphases on social references, environmental issues and opportunities to discover the practical impact of science in everyday life (EC, 2011). Therefore, part of hypothesis 1b assuming that several expectations of desirable science education expressed by the participants of this study relate to recommendations from literature can be confirmed for the third round as well.

The general assessments of the three concepts by the different sub-sample groups mainly reflect the tendencies of the general assessment of the total sample and in most cases, the concepts do not differ significantly from each other in their assessments. A differentiated view on the sub-samples shows, that on the basis of the average assessments of all three concepts in the different sub-samples, the highest priorities are in the same sequence as in the priority assessments of the second round given by the group of science education researchers (M=5.1), followed by the teachers (M=4.8) and scientists (M=4.6). The lowest priorities are provided by the students (M=4.2). Most similar priorities are given by the scientists, while education researchers differentiate to a higher degree between their assessments than the other groups.

Further comparisons of the different sub-samples show that although in general, they assess the concepts with similar priorities, they differ in a number of pair comparisons from each other. However, the sub-samples differ significantly only in the minority of cases (38%), so that hypothesis 5b regarding differences between the sub-samples in their specific assessments of the concepts cannot fully be confirmed in case of the priorities. In contrast, when comparing the students and the group of adult sub-samples, a tendency towards higher overall priorities can be found in the group of adult sub-samples (M=4.8). Moreover, significant differences between these two groups in 2 out of 3 concepts confirm the assumption of a particular gap between students' and adults' expectations (hypothesis 5c)[32]. It can be noticed that in the group of adult sub-samples, Concept A is assessed very similarly or even equally to Concept C. This implies that according to the group of adult sub-samples, contexts with current, social, globally relevant and occupational references as well as contexts of nature, everyday life and living environment should both play important roles in science education.

A more differentiated view on the concepts' realization in practice shows that the highest overall realization of the concepts is perceived by the teachers (M=3.3), followed by the education researchers (M=3.2) and students (M=3.1). The lowest overall realization of the three concepts in practice is seen by the scientists (M=2.7). Most similar practice assessments of the three concepts can be found in the group of scientists (SD=0.14), whereas in the same way as in the priority assessments, education researchers differentiate to a higher degree between their assessments than the other groups. All in all, the concepts are seen as realized to very similar degrees. Also, as no differences among the pair-comparisons of groups appear, the sub-samples share noticeable agreement about their views on the realization of the concepts. Hence, hypothesis 5b proposing particular differences in the specific opinions of the stakeholder groups cannot be confirmed for the practice assessments. A comparison of the students and the group of adult sub-samples shows that the opinions on the realization of the concepts mostly match so that hypothesis 5c regarding a particular gap between the young generation and the adult groups' views on the concepts' realization in science education practice cannot be confirmed either. All in all, the practice assessments of the different sub-sample groups provide a very homogeneous picture of all three concepts being perceived as not very much realized in science education.

With high priorities and rather low practice assessments of the three concepts, large gaps between the concepts' priority and realization and thus considerable dissatisfaction with the reality of science education compared with the priorities of the concepts can be identified in all sub-samples. The concepts' priority-practice differences in the different sub-samples indicate similar dissatisfaction.

In the same way as in the total sample, the largest gaps occur in the three adult sub-samples for the concept referring to general science-related education (Concept C), closely followed by the concept related to awareness of the scienc-

---

32 However, it should be noted that in cases of differences, the assessments of the groups do not contradict each other but feature similar tendencies within their emphases.

es in different contexts (Concept A). The students express higher dissatisfaction for Concept A than for Concept C. The strongest overall dissatisfaction with the concepts' realization in practice in general is expressed by both science education researchers and scientists (M=1.9). The students (M=1.1) show in the cases of all concepts comparably less dissatisfaction with the current state of science education than the other sub-samples. Yet, the sub-samples show mostly similar dissatisfaction with the concepts' realization. In this regard, hypothesis 5a proposing differences in the particular views of the stakeholder groups cannot be confirmed concerning the extent of dissatisfaction. As the adult sub-samples – while showing stronger dissatisfaction – only differ significantly from the students in the priority-practice difference of Concept A, hypothesis 5b proposing a particular gap between the young generation and adult stakeholders can not be confirmed with respect to dissatisfaction either. All in all, it can thus be said that the sub-samples mostly agree on underlining the high importance of the given concepts and on their insufficient presence in current science education.

With regard to the issue of the sample distribution in the third round being not equally balanced with respect to the four sub-samples – as can be expected in Delphi studies due to differing response rates – the very small deviations between original and normalized values show that the results are not misrepresented by the uneven distribution of the sample.

Assessment According to Different Educational Levels

A further question that arose from the findings of the second round was whether all of the three concepts should receive the same emphasis at all educational levels and how far these concepts are adequately realized at the different educational levels. The results of the analysis of the concepts' assessments according to different levels of education show that for all three concepts, both the importance and the extent of realization increases with the level of education from pre-school over elementary education and lower secondary education to higher secondary education. Hence, hypothesis 2c, suggesting a dependency of the concepts' priorities on the level of education in terms of an increase with advancing levels of education can be confirmed. As with the general priority assessments, Concept B is given lowest importance at all levels of education. The highest priority mean values for science education at pre-school and elementary level appear for Concept C. For lower secondary education, Concept C and Concept A are assessed at equally high levels and for higher secondary education, Concept A slightly surpasses Concept C. While all concepts are considered as not so important for pre-school and elementary education, the importance increases with advancing levels of education to high priority at higher secondary education. Also, while the concepts differ in their attributed importance at pre-school and elementary level, their priorities converge with the level of education and are considered with similarly high priority at higher secondary education.

As with the priority assessments, the perceived realization of all three concepts increases with the level of education. Thus, hypothesis 3c proposing the extent of realization depending on the level of education with an increase of the realization with advancing levels of education can be confirmed. This finding shows that the increasing importance of the concepts with advancing levels of education is perceived as taken into account in practice. Yet, none of the concepts is at any of the levels seen as more than fairly realized. Similarly to the trend in the priority assessments, the practice assessments converge with the level of education towards all three concepts being realized to similar extents at higher secondary education. Concept B is assessed with lowest realization at all levels of education. Concept C features the highest practice mean values for science education at pre-school, elementary level, and lower secondary level. For higher secondary education, the three concepts' realization is assessed at almost the same levels. At all levels of science education, Concept A and Concept B are assessed as present to very similar degrees.

With respect to the priority-practice differences of the concepts at different levels of education, it can be noticed that while the increasing importance of the concepts seems to be taken into account in practice, the extent of realization is smaller than the importance for all concepts at all four educational levels. This shows an underrepresentation of the concepts in practice with regard to their given priorities and suggests that from the perspective of the stakeholders involved in this study, science education does at none of the educational levels realize the given concepts to an adequate degree. Hence, the priority-practice differences show that although scientific literacy is widely accepted as the main aim of science education, the three identified concepts of scientific literacy based science education considered as relevant and important are from the participants' point of view not realized in accordance to their priorities at the different education levels. This suggests that although several priorities of the stakeholders can be compared to recommendations at both national (KMK, 2005a, 2005b, 2005c; MNU, 2003) and European level (e.g. Eurydice, 2011), a lot of highly relevant aspects are not seen as reflected in the reality of science education.

The priority and practice assessments of the concepts drift most strongly apart at lower and higher secondary education. This shows that particularly strong dissatisfaction with the realization of the concepts at higher levels of education can be identified. Comparing the concepts' priority-practice differences, Concept B appears as the least deficient concept at all educational levels. Concept C is seen to be most underrepresented at pre-school and elementary level, whereas Concept A is perceived as more underrepresented in lower and higher secondary education. While Concept B is characterized by scientific thinking, content knowledge and interdisciplinary scientific contexts, Concept A relates to a variety of contexts with current, social, globally relevant and occupational references, and Concept C refers to contexts of nature, everyday life and living environment. This shows that in general, the stakeholders see references to contexts beyond scientific contents for all levels of education within the frame of general education as more important for meaningful science education than intellectual education in interdisciplinary scientific contexts. As intellectual education in interdisciplinary scientific contexts is at all educational levels seen as less underrepresented than the other concepts, Concept B is in the least need of additional emphasis.

Moreover, the differentiated results show that according to the stakeholders included in this study, general science-related education and facilitation of interest in contexts of nature, everyday life and living environment (Concept C) should receive a stronger focus in basic education (pre-school and elementary level), while more advanced science education (lower and higher secondary education) should place more emphasis on aspects related to an awareness of the sciences in current, social, globally relevant and occupational contexts relevant in both educational and out-of-school settings (Concept A).

5 Final Discussion

After the results were described and discussed with regard to the research questions and hypotheses in the previous parts according to the three rounds, I will summarize these discussions in the following according to the three rounds of investigation and reflect on the implications of the results in terms of their significance for future science education in Germany. For a larger picture, I will also relate the outcomes of this study to other national curricular Delphi studies conducted in the context of the PORFILES project and point out implications for science education in Europe. This section will be followed by a critical reflection of issues related to the applied method and an outlook.

5.1 Overall Discussion and Reflection of the Results

First Round

In the first round of this study, 193 stakeholders involved with science and science education shared their views on aspects of desirable science education in a comprehensive and multi-facetted way. As expected, none of the contacted stakeholders in the group of education policy took part in the Berlin Curricular Delphi Study in Science. As all political parties and local administrative actors usually claim their demands in education topics, this result speaks volumes.

The participants' open-text responses provided first differentiated insights into expectations towards meaningful scientific literacy based science education. Following the procedure of statement analysis according to Bolte (2003b) and based on classifications from previous subject-specific curricular Delphi studies in physics, biology, and chemistry, as well as from research literature, a systematization of the participants' statements was achieved through establishing a category system. In order to process the statements as differentiated as necessary and as summarized as possible, the analysis of the stakeholders' responses led to a category system of 88 categories. Based on the structure of the questionnaire, the classification system illustrates that a broad variety of aspects related to contexts, motives, situations, concepts, topics, fields, perspectives, and qualifications are considered important for science education in Germany. This underlines that science education is not determined by single features, but shaped by a variety of factors. With an inter-rater reliability of 0.77, the procedure of qualitative content analysis complies with requirements of objectivity. In this way, the suitability of the established classification system for displaying the participants' responses adequately is confirmed. Therefore, as proposed at the beginning of this research endeavor, it was possible to establish a suitable classification system through applying the Delphi method for the purpose of representing the participants' opinions in the first round and as a basis for the quantitative analyses in the following round. Many of the identified categories are comparable to categories established in previous curricular Delphi studies in chemistry, physics and biology (Bolte, 2003b; Häußler et al., 1980; J. Mayer, 1992), and comply with recommendations about meaningful science education in literature (EC, 2011; KMK, 2005a, 2005b, 2005c; MNU, 2003; Schaefer, 2010b). Yet, the identified categories supplement these recommendations by several aspects as well. As the identified categories of the classification system are in the sense of statement bundles not mutually exclusive, it is important to underline that the identified categories of the classification system are not to be understood as separate building blocks on the basis of which scientific literacy can be put together. Rather, the categories provide first insights into what aspects of science education are mentioned by a heterogeneous sample of stakeholders as meaningful for scientific literacy based science education.

On the basis of the applied evaluation procedure in the first round, both first general tendencies as well as specific emphases can be identified in the opinions of the participants. In general, the mentioned aspects are of rather interdisciplinary nature and touch a variety of social, global and environmental concerns, public issues, and everyday life matters as well as fundamental scientific topics. Surprising within the outcomes of the descriptive statistical analyses of the responses are not so much the resulting high emphases about which general consensus is expressed (e.g. science in contexts of everyday life, media, current issues, society, global references, environment, medicine, scientific inquiry and content knowledge, and skills related to overarching aims of general education such as reflected and responsible acting, judgement, opinion-forming, comprehension, applying knowledge, and critical questioning). However, the low reference to those aspects mentioned relatively rarely by the experts were not expected (e.g. aspects related to knowledge about scientific occupations, basic scientific concepts and scientific perspectives). All the more interesting is thus the question whether the categories mentioned only rarely are generally considered less relevant or not taken into account by the participants in the first round for other reasons. Moreover, the question arises at this point if those aspects mentioned very often are actually realized to a high extent in educational practice. As proposed by Bolte (2003b), categories that are rarely mentioned might also play only minor role in the classroom so that representatives of sub-samples such as students or scientists were not aware of these aspects and thus not able to take these categories into account in their open-text responses of the first round. Insights regarding these issues are provided through the results of the second round of this study.

A closer look at the sub-samples shows that the science education research-ers and teachers feature the most differentiated responses. Also, these groups provided the most elaborate and sophisticated statements, which are likely to be driven by both their teaching experiences in science education and theoretical knowledge about science teaching. Their responses can be characterized by a focus on sustainability and citizenship. Moreover, their responses are strongly shaped by aspects that are often referred to as "scientific attitude", "higher-order thinking skills", "quantitative reasoning", "estimation skills", "problem-solving skills", and "decision-making skills". The scientists' responses can be described as the most pragmatic approaches and with strong references to scientific view-points, which might be explained by their background and occupation within areas of scientific research or in science-related professions in industry. The students' responses can be characterized as affected by aspects from their per-sonal environment, their everyday lives and media coverage.

While the sub-samples in the first round broadly correspond to each other in their general tendencies in terms of content, a more detailed comparison of the category frequencies shows that the sub-samples differ in their specific focuses and thus deviate from each other in 43% of the categories. Thus, on the basis of the results of the first round, no overall consensus among the different groups about aspects of desirable science education can be proposed. Comparing the group of students against the group of adult sub-samples, only in a minority of categories (19%), certain similarities seem to be shared by the adult sub-samples as opposed to the students. Therefore, the first round does not point towards a general gap between students and stakeholders from the adult sub-samples. Yet, in several categories, differences between the generations in their opinions of desirable science education become appear. Thus, based on the views of the participating stakeholders expressed in the first round, it can be stated that in several cases the expectations of the adult stakeholders do not coincide with the science educational interests of students, on whom education should essentially focus. Possible reasons for the differences between the generations might be related to different ways of approaching the question of desirable science educa-tion. While the adults, especially the science education researchers, might argu-ment from a more theory based perspective due to their professional back-grounds, the students might provide answers from a more classroom based point of view embedded in their own educational experiences. In this way, differences between students and adult stakeholders might as well be related to certain categories playing an only of minor role the classroom so that without explicit references, the students, in contrast to representatives of the adult sub-samples, were not aware of these aspects and consequently not able to take these categories into account in their open-text responses of the first round.

Second Round

As in the second round almost all given categories were assessed as important by the participants, the categories derived from the participants' responses in the first round can be seen as relevant criteria with respect to representing essential aspects of desirable science education. This can be seen as an indication that through the established classification system, the data from the first round could be analyzed not only with sufficient objectivity, but also with the necessary content validity. In this way, the applicability of the established category system is further confirmed. As such confirmation can only be reached through iteration – as part of the Delphi method – this is one of the findings that particularly result from the strength of the applied method.

In contrast to findings by Bolte (2003a), the relation between science and everyday life is not considered the most important aspect in the second round of the Berlin Curricular Delphi Study in Science but is found in the upper mid-range. Instead, while not contradicting findings by Bolte (2003a), the analyses of the priorities in the second round reveal that aspects related to the structure of the discipline and scientific topics are less important than aspects corresponding to the broad spectrum of both general and scientific inquiry related skills of scientific literacy and aspects that can be assigned to more overarching aims of education are considered as most important. Particular importance appears for aspects that refer to general skills and overarching competences, such as comprehension, analyzing, drawing conclusions, applying knowledge, creative and abstract thinking, judgement, opinion-forming, reflection, critical questioning, motivation, interest, and acting in a reflected and responsible way, as well as to elements that can be assigned to scientific inquiry such as self-dependent, structured, and precise working, awareness, and observation.

With almost all given categories in the second round being assessed as important by the participants, it can also be said that most of the categories that were relatively rarely considered by the participants in the first round of this study do not necessarily reflect low priorities but are indeed considered relevant. Thus, the results of this study show that essential aspects of desirable and meaningful science education cannot be identified by open questions only (as done in the first round), but it is necessary that the participants elaborate further on this issue by assessing categories that were systematically identified on the basis of their responses (as done in the second round). The assumption that low consideration of certain categories might be related to the corresponding aspects playing an only minor role in the classroom is supported by the analyses of the practice assessments: The results of the practice assessments show that although the majority of the categories are seen as relevant and important for modern science education, only few of these aspects are according to the participants adequately addressed and realized to sufficient degrees in the reality of science education. This might explain why in the first round certain aspects were not addressed to the extent that would have been expected in light of recommendations for science education in literature.

Moreover, the descriptive statistical results of the practice assessments reveal that science education within the frame of general education is perceived as mostly dominated by aspects related to scientific concepts, the traditional division of the science subjects, content knowledge, and scientific disciplines as well as specialized sub-disciplines. Issues that reach beyond the systematizations of the science subjects and an approach towards science from a more general education perspective are perceived as playing only a minor role in science education practice. These results are similar to the findings in the curricular Delphi study in chemistry, which reveal highest realization in practice for aspects related to terminology, basic ideas, content knowledge, and scientific concepts (Bolte, 2003a). Certain similarities can also be found with the outcomes of a study by Osborne and Collins (2000), which reveals students and parents' opinions of science education as an experience dominated by scientific content, lacking opportunities for discussion as well as a fragmented endeavor that leaves students without sufficient overview. That subject-specific aspects play an important role in science education is not surprising; after all, science education is about science. However, that these subject-specific aspects dominate science education to such an extent that overarching aims of general education are pushed aside is a notable finding of this study.

The results of the priority-practice differences in the second round bridge the gap to the question of which areas of science education are considered as especially deficient with respect the categories' representation in practice. In general, 65 of 80 categories, which equals 81%, show differences of M>0.75. This extent of differences points to a large underrepresentation of the collected aspects, indicating that current science education does not live up to the expectations of the stakeholders involved in this study. These findings lead to the conclusion that there is remarkable discontent with the current practice of science education in Germany. The only aspects with overall or sub-sample specific overrepresentation refer to curriculum framework as well as to scientific sub-disciplines and scientific concepts, which can be seen as an indication that science education is perceived as too much focused on the structure of the disciplines and specialized content. Relatively small, while still pointing to underrepresentation, are only those priority-practice differences of categories related to certain specialized fields, scientific sub-disciplines and scientific concepts. Areas of science education with most need for attention and improvement can be identified by their large priority-practice differences. The largest gaps between priority and realization point towards areas that are particularly underrepresented in science education and thus seen to be in most urgent need for improvement. The highest dissatisfactions appear for those areas that were at the same time considered as most important.

With respect to implications for science educational practice, the results indicate that in general, science education should place stronger emphases on enhancing skills and competences related to more general aims of education, address the relation between science and society, take more interdisciplinary approaches into account and should include stronger student orientation. Also, it can be concluded from the stakeholders' opinions in this study that deficits in current science education could be addressed by placing higher emphasis on fostering the students' abilities of judgement, opinion-forming, and reflection. Also, great importance should be placed on establishing educational offers that are able to foster the students' motivation and interest. A primary demand is thus to find ways in practice that make it possible to promote goals of science related general education in more comprehensive ways, in particular the goals of enhancing abilities of acting in a reflected and responsible way, applying knowledge, analyzing, drawing conclusions, and working in a self-dependent, structured, and precise manner. In addition, the statistical analyses show that particular attention should be paid to not merely passing on information and letting students memorize a series of unconnected facts, but making sure that the addressed issues are also understood by the students. Moreover, science education should focus more strongly on interdisciplinary themes, especially in contexts of society, nature, current issues, everyday life, education, and health, to include references to current scientific and interdisciplinary research and the nature of science, to foster skills in the context of scientific inquiry, to address consequences of developments resulting from scientific technical progress, and to discuss science related topics also in the context of different values. Special emphasis should also be placed on fostering the students' abilities of critical questioning. This implication should be highlighted because the results of the analyses show that it is especially this goal of science related general education that remains according to the stakeholders of this study largely unmatched. The results moreover make clear that science education mainly focusing on scientific contents does not comply with demands of contributing as well to the students' general education in the sense of scientific literacy.

Comparing the sub-samples, the highest mismatch between priority and re-alization of the given aspects and thus greatest dissatisfaction with the current status of science education is expressed by the group of science education re-searchers (M=1.5), as they gave the highest priorities and at the same time saw lowest realization of the aspects in practice. As the students consider all given aspects on average as comparably more represented in the classroom than the other sub-samples and at the same time gave lower priorities, this group express-es in contrast the lowest dissatisfaction with current science education (M=0.8). These tendencies are also illustrated by the individual priority-practice differ-ences. In the group of students, only 46 of 80 categories (58%) show priority-practice differences with mean values of M>0.75 . In contrast, differences with M>0.75 appear in the group of teachers for 63 of 80 categories (79%), in the group of education researchers for 71 categories (89%), in the group of scientists for 68 categories (85%), and in the adult sub-samples altogether for 69 categories (86%). Yet, a statistical comparison among the sub-samples reveals a total of 130 significant differences out of 480 possible pair comparisons of the priority-practice differences (27%). Hence, in contrast to anticipated differences among the sub-samples in their specific assessments of desirable science education, the majority of pair comparisons show a consensus among the sub-samples regard-ing their extent of dissatisfaction with the imbalance between desirable science education and its current state. Yet, the sub-samples still differ in their dissatis-faction in almost one third of the given aspects. This finding points to several underlying distinctions among the sub-samples with regard to their views on improvement of science education. Therefore, the frequently invoked consensus on scientific literacy based science education might still be questioned and the importance of including all affected stakeholders in a reflection on desirable science education is further underlined. It should be noted that the assessments of the sub-samples do not contradict each other such cases of significant differ-ences, but feature different accentuations within their emphases. The highest agreement on aspects which are most in need of action can be found between teachers and education researchers. This finding could be related to these two groups' professionally based similar perspectives on science education. Howev-er, a verification of this assumption goes beyond the scope of this study and would require further investigations. The highest number of significant differ-ences (almost 50% of the categories) can be found between students and educa-tion researchers. This result reiterates tendencies from the first round and shows again that in several aspects, especially students and education researchers seem to not share the same opinions on meaningful science education.

When comparing the students with the group of adult sub-samples, significant differences between the priority-practice differences can be found in almost one third of the categories. Therefore, it can be concluded that students and adults affected by science education share for the majority of categories similar dissatisfaction. Hence, no general gap between the generations can be proposed on the basis of the outcomes of the second round. Yet, the proportion of deviating emphases points towards different accentuations in these two groups' especially demands for improvement concerning scientific literacy based science education in terms of a stronger integration of students' views into current discourse on improving science education.

Addressing the question how far the views of the different sub-samples have changed or converged between the first and the second round in view of the general opinion as represented by the category system, it can be said that in contrast to the first round with the four sub-samples expressing different emphases in 43% of the categories, the proportion of significant differences between the sub-samples have diminished to 24% in the priority assessments, 29% in the practice assessments, and 27% in the priority-practice differences in the second round. Therefore, a certain convergence in the opinions of the sub-samples has taken place in the second round. This process shows that no preliminary conclusions should be drawn from such previous distinctions.

An interesting finding with regard to a comparison of all adult stakeholders with the group of students is that while these two groups generally agree in the majority of cases, the proportion of differences between these two groups has clearly increased from 19% in the first round to 31% (priority), 38% (practice), and 36% (priority-practice differences) in the second round. Such findings are only possible by conducting a second round and thus illustrate an important advantage of the Delphi method. This further validates the choice of the applied method.

Third Round

In the third round, all three concepts of desirable science education identified in the second round by means of hierarchical cluster analysis were assessed with rather high or high priority. This indicates that all three concepts are seen as important for science education and that by applying the Delphi method, a synopsis of the stakeholders' opinions on desirable science education in terms of identifying concepts could be achieved. Highest importance was given to the general and everyday life related concept of science education (Concept C), followed by the concept related to awareness of the sciences in different contexts (Concept A) and the concept referring to intellectual education in interdisciplinary scientific contexts (Concept B). In contrast to the priority assessments the concepts are considered by the participants as not very present in science education in general. Lowest realization is expressed for Concept A. Concept C is considered as most realized, yet only fairly present in practice.

Again, large priority-practice differences reveal that all three concepts fall short of their given priority for science education in general and that the demands posed by the participating stakeholders regarding scientific literacy based science education are not fulfilled in practice. The concept considered as most deficient in reality and thus providing contexts of most need for improvement is Concept A. In contrast, Concept B is, while still underrepresented, seen as the most adequately represented of the three concepts in practice, as already indicated by the results of the second round of this study, which reveal a lot of aspects included in Concept B being perceived as more adequately realized in current science education than other aspects. While Concept B places an emphasis on content knowledge and intellectual education in interdisciplinary scientific contexts, concepts A and C refer to a large variety of contexts, topics and skills with current, social, globally relevant and occupational references (Concept A) as well as aspects in contexts of nature, everyday life and living environment (Concept C).

When comparing the assessments of the adult sub-samples with the students, students see the concepts more adequately realized in practice (M=1.1), while the largest priority-practice differences and thus highest dissatisfaction with science education in terms of the realization of the concepts, is expressed in the groups of science education researchers and scientists (M=1.9). This finding is similar to the outcomes of the second round. Again, the four sub-samples differ in some pair comparisons from each other. Yet, these differences account only for a minority of cases (38%), so that they show overall similar dissatisfaction with the concepts' realization and no general disagreement among the sub-samples regarding improvement in the context of the given concepts can be proposed. When contrasting the students against the group of adult sub-samples, it becomes apparent that the adult stakeholders – while showing generally larger priority-practice differences – only differ significantly from the students in their dissatisfaction with the realization of Concept A. Hence, on the basis of the concepts' assessments in the third round, no particular gap between the generations can be proposed.

With respect to the question of how far the views of the different sub-samples have changed or converged between the second and third round in view of the general opinion, which was represented through the identified concepts of desirable science education, no further convergence has taken place between the sub-samples. Instead, similar to the second round, partial disagreement between the sub-samples seems to remain in the third round. Comparing students and the three adult-sub-samples, considerably higher disagreement than in the second round appears for the priorities (67%), while the disagreement about the priority-practice differences of the concepts has decreased from 36% to 17%. This shows that within the frame of a more condensed picture of the participants' opinions, the young and adult generation mostly share the same views on the need of improvement in science education. Apparently, in terms of demanded improvement, there is greater consensus for the given concepts than when offering several individual aspects. Again, such findings are only possible by applying a third round and underline the strength of the applied method.

A central finding from the differentiated assessments of the concepts in the third round is that for all three concepts, both the importance and the extent of realization increases with the level of education from pre-school and elementary education to lower secondary education and higher secondary education. However, the mismatches between priority and practice are larger with more advanced levels of education. For all levels of education, the stakeholders see references to contexts beyond scientific contents within the frame of general education (as expressed by Concept A and Concept C) as more important for meaningful science education than intellectual education in interdisciplinary scientific contexts (as expressed by Concept B). At the same time, based on the priority-practice differences, general science-related education and facilitation of interest in contexts of nature, everyday life and living environment (Concept C) should receive a stronger focus in basic education (pre-school and elementary level), while more advanced science education (lower and higher secondary education) should place more emphasis on aspects related to an awareness of the sciences in current, social, globally relevant and occupational contexts relevant in both educational and out-of-school settings (Concept A). With the smallest priority-practice differences on all levels of education, intellectual education in interdisciplinary scientific contexts (Concept B) is considered as more adequately realized in science education than the other two concepts and thus in the least need of additional emphasis. Hence, from the perspective of the stakeholders involved in this study, it can be said that science education with references to a variety of contexts is seen as more important than traditional scientific contents. This seems to imply that content knowledge should be subservient to more general issue. This conclusion can be compared to claims in several European countries for more practice-related and context-based science education with emphases on social references, environmental issues and opportunities to discover the practical impact of science in everyday life (Eurydice, 2011). However, this does not rule out content knowledge as an important focus within a scientific literacy oriented science curriculum. Instead, as all of the three concepts are perceived as underrepresented, it can be argued that especially the combination of the three concepts would account for desirable and meaningful scientific literacy based science education.

Sample

With an initial sample of 193 stakeholders in the first round and 154 remaining stakeholders in the second round, a total of 109 stakeholders have participated in this study throughout all three rounds. For a long-term multi-round study including qualitative design approaches, this is a very satisfying number and an acceptable regression in the sample size. In contrast to other Delphi studies conducted in the field of science education with smaller samples sizes, the size and solid composition of the sample in this study allowed a closer look at the four sub-samples (students, science teachers, science education researchers, and scientists), as special emphasis was placed on sufficient and balanced representation of each stakeholder group on sub-sample level when compiling the sample.

One of the main methodological strengths of this study is to show that essential aspects of desirable science education cannot be identified by open questions only (as done in the first round). In order to gain more precise findings, it was necessary that the categories identified from the participants' responses were assessed in more detail (as realized in the second round). To account for the large variety of opinions and to obtain a more comprehensive picture of the stakeholders' views, data reduction through which the results from the first round are condensed was crucial (as performed in the second round as well). By presenting the results in a more summarized format to the stakeholders for further assessment (as carried out in the third round), it was possible to gain a synopsis and thus further insights of the stakeholders' views in terms of empirically identified concepts of desirable science education. In this way, this work goes beyond an instantaneous picture in terms of a single survey and succeeded in describing and reconstructing expert opinions on desirable science education in a long-term study on a scientifically and empirically solid basis. For this purpose, the chosen method proved to be applicable. This finding should be particularly highlighted, as previous considerations and elaborations in the context of scientific literacy based science education usually applied more theory-based and more descriptive than research-based approaches. Also, this finding endorses the notion that complex issues in science education cannot be addressed by single surveys only.

Conclusion

Although scientific literacy has become an internationally well-recognized contemporary goal of science education in schools, the precise meaning of scientific literacy and the extent of consensus about the practical implications of adopting it as a central aim of science education in schools is often uncertain (Laugksch, 2000, p. 71). Moreover, many voices of affected stakeholders in society usually remain unheard on this issue (cf. 3.2). This study attempts to reduce this lack of research based evidence.

The particular strength of this work is a comprehensive scientific analysis of different German stakeholders' views of desirable scientific literacy based science education on the basis of the Delphi method. In this way, this work provides a scientifically and empirically profound approach towards a modern understanding of scientific literacy. As a result, the findings offer an orientation framework and meaningful starting points regarding the question how science education could be improved to better fulfill its role to support young people in becoming scientifically literate citizens. As previous Delphi studies in the field of science education focused on individual subjects or special aspects of science education only, the results obtained within the frame of this study can on the basis of an integrated approach to science be considered as especially significant.

In particular, the results might serve as a basis for inspiration in the context of improving science education by providing recommendations for enhancing the development of learning and teaching materials that aim at fostering scientific literacy or for the preparation of professional development programs for teachers, facilitating the uptake of innovative science teaching and the enhancement of scientific literacy, as administered within the PROFILES project (Bolte et al., 2014; Bolte, Holbrook, et al., 2012; Bolte & Rauch, 2014).

If a realization of the recommendations that can be derived from the results of this study with the aim of contributing to the students' general education in terms of scientific literacy is taken into account seriously, different emphases in science education than those of current practice have to be applied. It can be assumed that such a shift would not occur at the expense of learning outcomes in science, as PISA (OECD, 2004a, 2007b) and TIMSS (Martin et al., 2012) show that even the current format of conventional science education, which strongly focuses on scientific content, does not lead to the demanded subject-specific learning outcomes, let alone that the current format of science education contributes to motivating the students to engage with science in contexts beyond school or, in the sense of lifelong learning, after their graduation (Bolte, 2003a, p. 39). In this sense, a consistently applied reform of science education is overdue.

What moreover endorses the need for a revision of science education according to the recommendations derived from the results of this study is the demand of taking into account the interests and needs of students. On the basis of the different perspectives covered by the sample of this study, it is possible to develop empirically based learning environments that take into account the views of the students as well. Following Bolte (2003a), and on the basis of find-ings in the fields of pedagogical interest theory (Gräber, 1995; Prenzel, 1988), constructivist learning theories (Glasersfeld, 1993, 1996), and *Bildungsgangdid-aktik* (Meyer, 2005; Meyer & Reinartz, 1998; Schenk, 2005) it can be assumed that such an approach would lead to an improvement of science education. In this way, the outcomes of this study provide a valuable basis for discussions about issues of scientific literacy based science education. On the basis of their practical implications, the results might also serve as a source for consulting in the field of curriculum development and education policy.

5.2 Comparison to Other National Curricular Delphi Studies Conducted in the Context of the PROFILES Project

Having analyzed German stakeholders' views about desirable science education, the question arises how the findings obtained throughout this study compare to the outcomes in other countries and whether they can be replicated on a Europe-an level. A consideration of this question is possible through the PROFILES project. Within the International PROFILES Curricular Delphi Study on Science Education in the context of Work package 3, which focuses on the involvement and interaction of stakeholders affected by science education, several curricular Delphi studies conducted by each participating institution provide insights into local stakeholders' opinions on their national science education. Within this context, it was possible to gather views from more than 2700 participants in 19 countries about aspects of science education that can be considered meaningful and pedagogically desirable for the individual in the society today and in the near future. Through several comparisons and meta-analyses, it was possible to gain insights about a European[33] perspective on the current status and possible defi-ciencies of science education (Gauckler, 2014b; Gauckler et al., 2014; Schulte, Bolte, et al., 2014; Schulte & Bolte, 2012; Schulte, Georgiu, et al., 2014).

On the basis of a comparison of the classification systems developed from their stakeholders' statements in the first round by 20 institutions from 19 differ-ent European or Europe associated countries with a total of 2 706 participants, it was found that despite differences in the number of categories and terminology, the stakeholders' statements from the different participating countries correspond to each other to a large extent. Hence, first implications about a certain consen-sus between stakeholders from the participating countries concerning aspects that

---

33 Among the participating countries, Georgia, Israel, Switzerland, and Turkey are not members of the EU, yet, they are still in various respects associated with the EU and therefore considered in terms of a European perspective as well.

are perceived as relevant for science education could be identified (Gauckler et al., 2014, p. 129; Schulte & Bolte, 2012, p. 50). Categories included by every partner institution of this comparison refer to motivation, medicine, health, everyday life and environment (Gauckler et al., 2014, p. 129). With respect to approaching a modern understanding of scientific literacy from the perspective of different stakeholders on a European level, these category systems provide first hints for core issues on the basis of international overlapping of categories.

According to a comparison of the priority assessments in the second round including 18 partner institutions from 17 countries with a total of 1867 participants, the most important aspects of science education from a European perspective refer to more general skills and competencies related to scientific thinking and reasoning, such as analyzing and interpreting data and observations, critical thinking, judgement, reflection, and acting responsibly, as well as to content knowledge. This finding is similar to the results of the Berlin curricular Delphi Study in Science presented in this work and suggests that from a European perspective, general skills and competencies should be based on, considered in interaction with, and be acquired within the context of basic scientific knowledge, whereas traditional sciences and scientific sub-disciplines have lower priority (Gauckler et al., 2014, p. 129). In contrast, most of the given aspects of desirable science education were considered as not even fairly realized in the science classroom. According to the stakeholders involved in the different national curricular Delphi studies, current science education in Europe consists mainly of curriculum orientation, scientific contents and concepts of scientific sub-disciplines. In fact, none of the most important categories could be found amongst the most present categories, which can be seen as a first hint that important aspects of science education are underrepresented in practice (Gauckler et al., 2014, p. 129). This tendency was also found in previous comparisons (Schulte, Bolte, et al., 2014, p. 191; Schulte, Georgiu, et al., 2014, p. 131). As European countries mostly share the same general goals of science education in terms of equipping students with competencies, knowledge, skills and attitudes that are deemed important by each country (EC, 2004, p. vii), it is not surprising that the priority assigned to individual categories is comparable between countries. This finding points towards similar tendencies in different countries' and thus common goals of European science education. In contrast, as the countries differ in their science education practice, for example in their curricula and organization, the assessment of realization varies more noticeably between the different countries (Gauckler et al., 2014, p. 129).

The misrepresentation between wish and reality is further illustrated by contrasting the aspects deemed most important for science education by European stakeholders with the ones perceived as most realized in practice on the basis of priority-practice differences, as shown by Gauckler et al. (2014, p. 130). From such analyses, aspects can be identified that are considered especially important but not sufficiently realized in the science classroom, providing in this way hints about areas of current science education practice throughout Europe which might require improvement. In general, the vast majority of aspects are considered more important than they are present in science education and are therefore underrepresented. While scientific concepts in sub-disciplines appear to be more adequately represented in terms of their priorities, general skills and competencies related to scientific thinking and reasoning as well as the discussion of current scientific topics, their impact on society, and the promotion and consideration of students' interests and motivation are strongly underrepresented in science education practice in Europe and can thus be identified as the most deficient areas. Again, similarities can be found with findings by Schulte et al. (2014, p. 185), who identified shortcomings for aspects related to the connection between science and everyday life, the promotion of students' interest in science, the realization of IBSE and other overarching educational goals. These areas can be related to what is defined by the European Commission (EC, 2007, p. 6) as one of the main goals of science education: To equip every young person with the skills necessary to live and work in tomorrow's society, which relies heavily on technological and scientific advances of increasing complexity.

In the third round of the International PROFILES Curricular Delphi Study on Science Education, stakeholders were asked to assess concepts of desirable science education. The concepts were developed within the Berlin Curricular Delphi Study in Science on the basis of category combinations that seemed meaningful to the stakeholders. They include "Awareness of the sciences in current, social, globally relevant and occupational contexts relevant in both educational and out-of-school settings" (Concept A), "Intellectual education in interdisciplinary scientific contexts" (Concept B), and "General science-related education and facilitation of interest in contexts of nature, everyday life and living environment" (Concept C). In order to investigate whether the European stakeholders involved in the International PROFILES Curricular Delphi Study on Science Education would consider these concepts relevant for science education, the concepts were assessed in terms of priority and practice not only in Germany, but in other participating countries as well. Due to drop-out and incomparability in some cases, the concepts' assessments collected by 11 institutions in 11 European countries[34] with a total of 1301 stakeholders were analyzed (Gauckler, 2014b, pp. 34–36). A comparison of the assessments in the third

---

34 Germany, Estonia, Austria, Czech Republic, Ireland, Latvia, Poland, Romania, Slovenia, Turkey and Georgia

round shows that in general, all three concepts were considered relevant for science education in Europe, as they were assessed with at least rather high priority by stakeholders from every country that was included in this comparison. Furthermore, as indicated in previous comparison as well already (Schulte & Bolte, 2014b), all three main findings from the third round of the Berlin Curricular Delphi Study in Science also apply to tendencies found in the European perspective: Firstly, the priorities given to the concepts are the higher the more advanced the educational level is. That means that the concepts become increasingly important from pre-school over basic to advanced science education. Secondly, the higher the educational level is, the more realized the concepts are considered to be in practice. Thirdly, the gaps between priority and extent of implementation become wider with increasing levels of education. Concept C was considered the most important concept for all levels of education. At the same time, it was seen as the most underrepresented concept at pre-school and elementary level. At lower and higher secondary education, Concept A was perceived as more underrepresented. Concept B, while important, was assessed with the lowest priority of all three concepts, although stakeholders from different countries assessed this issue very differently. As Concept B also showed the least underrepresentation on all levels of education, it can be proposed that

> "European stakeholders consider intellectual education in interdisciplinary scientific
> contexts as more appropriately covered in science education than the other two con-
> cepts" (Gauckler, 2014, p. 36).

Hence, general science-related education in contexts of nature and everyday life (Concept C) should be more stressed in basic science education (pre-school and elementary level). More advanced science education (lower and higher secondary school) should place stronger emphasis on aspects related to current, social, globally relevant and occupational contexts (Concept A). Intellectual education in interdisciplinary scientific contexts (Concept B) is less underrepresented than the other concepts and thus seems to need the least additional accentuation. As Concept B includes content knowledge, scientific thinking, and interdisciplinary scientific contexts, while the other two concepts refer to further contexts, such as current, social, globally relevant and occupational contexts (Concept A) and nature, everyday life and living environment (Concept C), it can be suggested that "from the perspective of European stakeholders, reference to a variety of contexts is more important than more traditional scientific contents" (Gauckler, 2014, p. 36). Another result from the comparison of the assessments in the third round is that the stakeholders, although being associated with and referring to different educational systems, displayed a relatively high degree of agreement about the importance of the three concepts (Gauckler, 2014, p. 36). This finding can be related to the claim from the European Commission (EC, 2004, p. vii) that all European countries share the same general purpose of science education. Therefore, "the results of this meta-analysis might serve as inspiration for further studies on science education on a European scale" (Gauckler, 2014, p. 36).

All in all, it can be said that through a systematic comparison of the categories in the PROFILES partners' classification systems developed on the basis of their stakeholders' statements in the first round, of the priority and practice assessments of these categories in the second round of the national Delphi Studies, and of the priority and practice assessments of concepts of desirable science education of the national Delphi Studies in the third round, it was possible to obtain more detailed insights into the importance of the different aspects and the current situation of science education in Europe and thus provide first empirically based insights into "a European perspective on current science education" (Gauckler, 2014, p. 36). In this way, the results could potentially serve as a basis for further science curriculum related discussions and inspiration both for individual teachers and school reformers Europe-wide. Moreover, on the basis of their comprehensive and large-scale scope, the findings might provide valuable starting points for teachers in Europe to guide their students to appreciate the "relevance of learning through 'science' for lifelong learning, responsible citizenry and for preparing for meaningful careers" (Bolte, Streller, et al., 2012, p. 35). Hence, in order to find meaningful answers to the broad central question about desirable aspects of scientific literacy based science education that lies at heart of the International PROFILES Curricular Delphi Study on Science Education, "it seems possible or even necessary to unite research from different countries on a European level" (Gauckler, 2014b, pp. 34–36).

5.3 Critical Reflection of the Applied Method and Outlook

For a discussion of the limitations of the analyses made in this study, I will critically reflect the applied method, describe issues related to the sample choice and composition, and point out suggestions for further research.

With respect to the applied Delphi technique of this study, a remarkable result was that the majority of categories that were relatively rarely referred to by the participants in the first round of this Delphi study and therefore initially interpreted as being considered less important, are actually regarded as very relevant when specifically assessed in terms of their priority for desirable and meaningful science education. This finding points to fundamental methodological problems of surveys (Bolte, 2003a, p. 36), which can only be discussed at this point but not solved. Face-to-face interviews involve the risk of opinion leadership that can contribute to a distortion of the results. In contrast, in anonymous surveys it cannot be ruled out that the questions will not be answered with the desired or necessary reflection efforts and thus not comprehensively enough. In addition, face-to-face surveys and open question formats require a certain limitation of the sample due to the elaborate and complex methods of content analysis. On the other hand, systematized and standardized surveys, while allowing larger samples, considerably limit the range of possible answers of the participants. Hence, there is no ideal methodological way. Yet, despite all possible limitations that might be inherent in this method, one main strength of the curricular Delphi method is the possibility to reflect the findings of a round in light of the previous round(s). For the presented curricular Delphi study, this implies that the results of the first round have to be relativized in certain areas. However, it does not mean that the results from the first round are unreliable, inadequate, or less valid. On the contrary, also the fact that participants included certain categories relatively rarely in their responses in the first round allowed conclusions that were confirmed by the results of the second round.

As the effectiveness of the Delphi method depends on the particular nature of its implementation, such as the number of iterations, type of feedback, type of task that it is applied to, size and constitution of groups, and characteristics of group members, the design decisions of this study have to be critically reflected as well. With drop-out rates of 20% after the first round and 29% after the second round, the regression was relatively low and thus very satisfying for such a long-term study. At the same time, the sample size of this Delphi study with still 109 participants returning for the third round can be considered comparatively large. Taking into account these findings and the fact that a satisfying condensation of the participants' views was reached in the final round of this study, following previous recommendations of applying three rounds has proven useful. Due to the elaborate and complex methods of data analysis, large time spans occurred between the different rounds. Shortening these spans, for example by more simultaneous analyses, might have led to even lower drop-out rates. Although special emphasis when compiling the sample of this study was placed on obtaining an approximately even distribution of the sub-samples (students, science teachers, science education researchers, and scientists) to allow for more detailed analyses, differing sub-sample sizes emerged. Such differences result from differing response rates in the sub-samples. However, a comparison of the obtained data to normalized data showed no substantial deviations. This implies that the results were not considerably distorted by differences in sub-sample size. In contrast to other Delphi studies conducted in the field of science education with smaller samples sizes, the validity of the sample in this study allowed an analysis of the four sub-samples.

In some cases, participants reported that they considered the tasks in this study too complex or formulated on a too sophisticated level. In particular, such feedback was received from students in lower secondary education. Hence, a possible solution to reduce, while not eliminate, such problems in future studies of this kind could be some differentiation of the tasks in terms of language according to the specific characteristics of the different groups involved. Yet, the sub-sample size in the group of students in this study shows that students from both lower and higher secondary level were able to deal with the given tasks. Therefore, it is not assumed that the formulation of the task in this study has systematically discouraged students' participation.

Further methodological improvement can be suggested with respect to the third round. In the second part of this round, the stakeholders provided their assessments with respect to different levels of education such as pre-school, elementary level, lower secondary level, and higher secondary level, while not representing distinguished experts for these specialized educational levels. Experts for these particular educational levels could be more strongly focused on in follow-up studies.

As pointed out by Gauckler (2014a) with respect to the comparison of the national curricular Delphi studies conducted within the context of the PROFILES project, some of the participating institutions collected during the first round their participants' statements with questionnaires that deviated from the question and answer format described in this study. However, as the aim of the first round was to collect a broad variety of aspects related to science education these deviations are of little consequence. More relevant for a critical reflection of this comparative analysis is the deviation from the Delphi method by some institutions who decided to use the classification system by the Berlin curricular Delphi Study in Science or an adaptation of it to gather the priority and practice assessments from their stakeholders, as opposed to feeding back the local results from the first round in order to give the participants the opportunity to reflect on their opinions based on the answers of the whole panel. However, as shown by Gauckler (2014a), the classification system developed within the first round of the Berlin curricular Delphi Study in Science is generally well suited to represent a wide range of stakeholders' responses in the first round and therefore could be used as a basis for the second round in all countries. In fact, it was stated that – although certain local characteristics could not have been captured – if the Berlin classification system had been used by all partners for the second round, comparisons would have been more practical.

Another central problem for the comparison between countries in the second round reported by Gauckler (2014a) refers to terminology. In many cases, working groups of the partner institutions used in the first round a different label for a category which could be assumed to denote the same ideas and therefore were considered comparable. It was not possible to refer back to original statements of participants, as the national curricular Delphi studies were conducted in the local languages. Therefore, inaccuracies might have been created through the translation of category labels from local languages into English. If it can be assumed that European stakeholders mostly share the general goal of developing students individually and socially, even if the exact competencies, knowledge, skills and attitudes that are necessary for this goal might be different in each country (EC 2007, vii), then it can be inferred that high standard deviations for the priority assessments of categories between the different countries might indicate misunderstandings about the concept behind the label. However, it might also be possible that in cases of high standard deviation, stakeholders showed actual disagreement about the importance of the concept. One way of decreasing the risk of misunderstanding would have been for every partner institution to provide a glossary with descriptions and examples of the category labels in their local language, as done in the Berlin Curricular Delphi Study in Science (Gauckler, 2014a).

As underlined by Gauckler (2014a), for further purposes of comparisons, it would be useful to condense the number of categories in the classification system developed within the Berlin Curricular Delphi Study in Science, which was used as a basis for the applied comparative analyses. In addition, a condensed classification system would reduce the paper work and could potentially raise the willingness of people to participate and would therefore prove useful for further studies, for example when including further European and international countries. Further condensation could be possible by using the collected data in order to statistically analyze the overlap between categories, for example through exploratory or confirmatory factor analysis.

The outcomes of the comparative analyses of the results provided by national curricular Delphi studies carried out within the International PROFILES Curricular Delphi Study on Science Education, as discussed by Gauckler (2014a), were referred to as a European perspective. However, one of the participating European countries, while associated with the EU, is not a member of the EU (Switzerland), and three further countries are located on the borders or outside of Europe (Georgia, Israel, and Turkey). In contrast, several European countries are not represented in the comparative analyses (Belgium, Bulgaria, Croatia, Denmark, Estonia, France, Greece, Hungary, Lithuania, Luxembourg, Malta, Slovakia and the United Kingdom). This restraint is closely linked to the fact that not all European countries are present in the PROFILES project. Therefore, the European perspective referred to as a result of the comparison of the different national curricular Delphi studies carried out within the PROFILES project could be complemented with the aforementioned remaining European countries in further studies in order to provide a more complete representation of European stakeholders' opinions about desirable science education. Moreover, this picture could be enhanced towards an even more international frame by adding perspectives from North and South American, African, Asian and Australian countries (Gauckler, 2014a).

References

AAAS. (1990). Science for All Americans. New York: Oxford University Press. Retrieved from http://www.project2061.org/publications/sfaa/online/sfaatoc.htm [21.6.2014]

AAAS. (1993). Benchmarks for Science Literacy. New York: Oxford University Press.

AAAS. (1997). Resources for Science Literacy: Professional Development. New York: Oxford University Press.

AAAS. (2001). Designs for Science Literacy. New York: Oxford University Press.

Aichholzer, G. (2002). Das ExpertInnen-Delphi. In A. Bogner, B. Littig, & W. Menz (Eds.), Das Experteninterview (pp. 133–153). Opladen: Leske und Budrich.

Aikenhead, G. S. (2003). Review of Research on Humanistic Perspectives in Science Curricula. Paper Presented at the ESERA Conference, Nordwijkerhoud, The Netherlands. Retrieved from http://www.usask.ca/education/profiles/aikenhead/webpage/ESERA_2.pdf [07.05.2014]

Ammon, U. (2009). Delphi-Befragung. In S. Kühl, P. Strodtholz, & A. Taffertshofer (Eds.), Handbuch Methoden der Organisationsforschung (pp. 458–476). Wiesbaden: Verlag für Sozialwissenschaften.

Asendorpf, J. (2007). Psychologie der Persönlichkeit. Heidelberg: Springer.

Ayton, P., Ferrell, W. R., & Stewart, T. R. (1999). Commentaries on "The Delphi technique as a forecasting tool: issues and analysis" by Rowe and Wright. International Journal of Forecasting, 15(4), 377–379.

Bardecki, M. J. (1984). Participants' response to the Delphi method: An attitudinal perspective. Technological Forecasting and Social Change, 25(3), 281–292. http://doi.org/10.1016/0040-1625(84)90006-4

Bauer, H. H. (1992). Scientific Literacy and the Myth of the Scientific Method. Chicago: University of Illinois Press.

Baumert, J., Bos, W., & Lehmann, R. (Eds.). (2000). TIMSS/III: Dritte Internationale Mathematik- und Naturwissenschaftsstudie. Mathematische und naturwissenschaftliche Grundbildung am Ende der Pflichtschulzeit. Opladen: Leske und Budrich.

Baumert, J., Klieme, E., Neubrand, M., Prenzel, M., Schiefele, U., Schneider, W., … Weiß, M. (1999). Internationales und nationales Rahmenkonzept für die Erfassung von naturwissenschaftlicher Grundbildung in PISA. Berlin: Max-Planck-Institut für Bildungsforschung.

Bayrhuber, H., Bögeholz, S., Elster, D., Hammann, M., Hößle, C., & Lücken, M. (2007). Biologie im Kontext – Ein Programm zur Kompetenzförderung durch Kontextorientierung im Biologieunterricht und zur Unterstützung von Lehrerprofessionalisierung. Der Mathematische Und Naturwissenschaftliche Unterricht, 60, 282–286.

Beaton, A., Martin, M. O., Mullis, I. V. S., Gonzalez, E. J., Smith, T., & Kelly, D. (1997). Science Achievement in the Middle School Years: IEA's Third International Mathematics and Science Study (TIMSS). Chestnut Hill, MA: TIMSS International Study Center, Boston College.

Beattie, H. (2012). Amplifying student voice: the missing link in school transformation. Management in Education, 26(3), 158–160.

Becker, D. (1974). Analyse der Delphi-Methode und Ansätze zu ihrer optimalen Gestaltung. Frankfurt am Main: Doctoral dissertation.

Benner, D. (2002). Die Struktur der Allgemeinbildung im Kerncurriculum moderner Bildungssysteme. Ein Vorschlag zur bildungstheoretischen Rahmung von PISA. Zeitschrift Für Pädagogik, 48(1), 68–90.

Bennett, J., Gräsel, C., Parchmann, I., & Waddington, D. (2005). Context-based and Conventional Approaches to Teaching Chemistry: Comparing teachers' views. International Journal of Science Education, 27(13), 1521–1547.

Bennett, J., & Lubben, F. (2006). Context-based Chemistry: The Salters approach. International Journal of Science Education, 28(9), 999–1015.

Bennett, J., Lubben, F., & Hogarth, S. (2007). Bringing science to life: A synthesis of the research evidence on the effects of context-based and STS approaches to science teaching. Science Education, 91(3), 347–370.

Blankertz, H. (1984). Bildung im Zeitalter der großen Industrie. Schroedel.

Bolger, F., Stranieri, A., Wright, G., & Yearwood, J. (2011). Does the Delphi process lead to increased accuracy in group-based judgmental forecasts or does it simply induce consensus amongst judgmental forecasters? Technological Forecasting and Social Change, 78(9), 1671–1680.

Bolger, F., & Wright, G. (2011). Improving the Delphi process: Lessons from social psychological research. Technological Forecasting and Social Change, 78(9), 1500–1513.

Bolte, C. (2000). Delphi-Studie Chemie: Orakel irreleitenden Inhalts oder Orientierungshilfe zur Bewältigung von Bildungsaufgaben? In R. Brechel (Ed.), Motivation und Interesse; Voraussetzung und Ziel des Unterrichts in Chemie und Physik. Zur Didaktik der Physik und Chemie. Probleme und Perspektiven (pp. 229–231). Alsbach: Leuchtturm-Verlag.

Bolte, C. (2001). Chemieunterricht und Allgemeinbildung – Projektskizze der curricularen Delphi-Studie Chemie. eWi-Report, 23, 78–81.

Bolte, C. (2002). Die curriculare Delphi-Studie Chemie: Allgemeinbildung und Chemieunterricht. Chemkon, 9(2), 86–90.

Bolte, C. (2003a). Chemiebezogene Bildung zwischen Wunsch und Wirklichkeit - Ausgewählte Ergebnisse aus dem zweiten Untersuchungsabschnitt der curricularen Delphi-Studie Chemie. Zeitschrift für Didaktik der Naturwissenschaften, 9, 27–42.

Bolte, C. (2003b). Konturen wünschenswerter chemiebezogener Bildung im Meinungsbild einer ausgewählten Öffentlichkeit - Methode und Konzeption der curricularen Delphi-Studie Chemie sowie Ergebnisse aus dem ersten Untersuchungsabschnitt. Zeitschrift für Didaktik der Naturwissenschaften, 9, 7–26.

Bolte, C. (2008). A Conceptual Framework for the Enhancement of Popularity and Relevance of Science Education for Scientific Literacy, based on Stakeholders' Views by

Means of a Curricular Delphi Study in Chemistry. Science Education International, 19(3), 331–350.

Bolte, C., Holbrook, J., Mamlok-Naaman, R., & Rauch, F. (Eds.). (2014). Science Teachers' Continuous Professional Development in Europe. Case Studies from the PRO-FILES Project. Klagenfurt: Alpen-Adria-Universität Klagenfurt.

Bolte, C., Holbrook, J., & Rauch, F. (Eds.). (2012). Inquiry-based Science Education in Europe: Reflections from the PROFILES Project. Berlin: Freie Universität Berlin.

Bolte, C., & Rauch, F. (Eds.). (2014). Enhancing Inquiry-based Science Education and Teachers' Continuous Professional Development in Europe: Insights and Reflections on the PROFILES Project and other Projects funded by the European Commission. Klagenfurt: Alpen-Adria-Universität Klagenfurt.

Bolte, C., & Schulte, T. (2014a). Stakeholders Involvement and Interaction in PROFILES. In C. Bolte & F. Rauch (Eds.), Enhancing Inquiry-based Science Education and Teachers' Continuous Professional Development in Europe: Insights and Reflections on the PROFILES Project and other Projects funded by the European Commission (pp. 34–39). Klagenfurt: Alpen-Adria-Universität Klagenfurt.

Bolte, C., & Schulte, T. (2014b). Stakeholders' Views on Science Education in Europe: Method and First Insights of the PROFILES International Curricular Delphi Study on Science Education. In C. P. Constantinou, N. Papadouris, & A. Hadjigeorgiou (Eds.), E-Book Proceedings of the ESERA 2013 Conference: Science Education Research For Evidence-based Teaching and Coherence in Learning. Part 8 (co-ed. M. Ossevoort & J. A. Nielsen) (pp. 131–142). Nicosia, Cyprus: European Science Education Research Association. Retrieved from http://www.esera.org/media/esera2013/Claus_Bolte_19Dec2013.pdf [12.06.2014]

Bolte, C., & Schulte, T. (2014c). Wünschenswerte naturwissenschaftliche Bildung im Meinungsbild ausgewählter Experten. Der Mathematische Und Naturwissenschaftliche Unterricht, 67(6), 370–376.

Bolte, C., Schulte, T., Kapanadze, M., & Slovinsky, E. (2012). Stakeholders' Views on Desirable Science Education in Georgia. In M. Kapanadze & I. Eilks (Eds.), Student Active Learning in Science (pp. 79–84). Tbilisi: Ilia State University Press.

Bolte, C., & Streller, S. (2013). Education through Science - Bildung durch Naturwissenschaften. In Inquiry-based learning - Forschendes Lernen. Gesellschaft für Didaktik der Chemie und Physik. Jahrestagung in Hannover 2012 (pp. 180–182). Kiel: IPN. Retrieved from http://www.gdcp.de/index.php/tagungsbaende/tagungsband-uebersicht/145-tagungsbaende/2013/4220-band33 [06.03.2013]

Bolte, C., Streller, S., Holbrook, J., Mamlok-Naaman, R., Hofstein, A., & Rauch, F. (2011). PROFILES – Professional reflection-oriented focus on inquiry-based learning. Proceedings of the European Science Educational Research Association (ESERA). Lyon. Retrieved from http://lsg.ucy.ac.cy/esera/e_book/base/ebook/strand5/ebook-esera2011_BOLTE_2-05.pdf [30.05.2012]

Bolte, C., Streller, S., Holbrook, J., Rannikmae, M., Hofstein, A., Mamlok-Naaman, R., & Rauch, F. (2012). Introduction into the PROFILES Project and its Philosophy. In C. Bolte, J. Holbrook, & F. Rauch (Eds.), Inquiry-based Science Education in Europe:

Reflections from the PROFILES Project (pp. 31–42). Berlin: Freie Universität Berlin.

Bonnekoh, W. (1992). Naturwissenschaft als Unterrichtsfach: Stellenwert und Didaktik des naturwissenschaftlichen Unterrichts zwischen 1800 und 1900. Frankfurt am Main: Lang.

Borg, I. (1992). Grundlagen und Ergebnisse der Facettentheorie (1. ed.). Bern: Huber.

Börlin, J., & Labudde, P. (2014). Swiss PROFILES Delphi Study: Implication for Future Developments in Science Education in Switzerland. In C. Bolte, J. Holbrook, R. Mamlok-Naaman, & F. Rauch (Eds.), Science Teachers' Continuous Professional Development in Europe. Case Studies from the PROFILES Project (pp. 48–58). Klagenfurt: Alpen-Adria-Universität Klagenfurt.

Bortz, J. (2005). Statistik für Human- und Sozialwissenschaften. Heidelberg: Springer.

Bortz, J., & Döring, N. (2006). Forschungsmethoden und Evaluation. Berlin:Springer.

Bos, W., Bonsen, M., Baumert, J., Prenzel, M., Selter, C., & Walther, G. (Eds.). (2008). TIMSS 2007. Mathematische und naturwissenschaftliche Kompetenzen von Grundschulkindern in Deutschland im internationalen Vergleich. Münster: Waxmann.

Brown, B. B. (1968). Delphi Process. A Methodology Used for the Elicitation of Opinions of Experts. Retrieved from http://www.rand.org/pubs/papers/P3925.html [12.11.2013]

Brown, B. B., Cochran, S. W., & Dalkey, N. C. (1969). The Delphi Method II: Structure of Experiments. Santa Monica: RAND Corporation. Retrieved from http://www.rand.org/pubs/research_memoranda/RM5957.html [12.11.2013]

Bünder, W. (1997). Practising Integration in Science Education: PING. In W. Gräber & C. Bolte (Eds.), Scientific Literacy (pp. 399–414). Kiel: IPN.

Bund-Länder Kommission für Bildungsplanung und Forschungsförderung. (1997). Gutachten zur Vorbereitung des Programms „Steigerung der Effizienz des mathematisch-naturwissenschaftlichen Unterrichts". Bonn: Bund-Länder Kommission für Bildungsplanung und Forschungsförderung.

Burkard, U., & Schecker, H. (2014). Curriculare Delphi Studien. In D. Krüger, I. Parchmann, & H. Schecker (Eds.), Methoden in der naturwissenschaftsdidaktischen Forschung (pp. 159–168). Berlin & Heidelberg: Springer.

Burns, T., O'Connor, J., & Stocklmayer, S. (2003). Science communication: a contemporary definition. Public Understanding of Science, 12, 183–202.

Bybee, R. W. (1997). Toward an Understanding of Scientific Literacy. In W. Gräber & C. Bolte (Eds.), Scientific Literacy (pp. 37–68). Kiel: IPN.

Bybee, R. W. (2002). Scientific Literacy - Mythos oder Realität? In W. Gräber, P. Nentwig, T. Koballa, & R. Evans (Eds.), Scientific Literacy. Der Beitrag der Naturwissenschaften zur allgemeinen Bildung (pp. 21–43). Opladen: Leske und Budrich.

Bybee, R. W. (2008). Scientific literacy, environmental issues, and PISA 2006: The 2008 Paul F- Brandwein Lecture. Journal of Science Education and Technology, 17, 566–585.

Bybee, R. W., Fensham, P., & Laurie, R. (2009). Scientific literacy and contexts in PISA 2006 science. Journal of Research in Science Teaching, 46(8), 862–864.

Bybee, R. W., & McCrae, B. (2011). Scientific Literacy and Student Attitudes: Perspectives from PISA 2006 science. International Journal of Science Education, 33(1), 7–26.

Bybee, R. W., McCrae, B., & Laurie, R. (2009). PISA 2006: An Assessment of Scientific Literacy. Journal of Research in Science Teaching, 46(8), 865–886.

Charro, E., Plaza, S., & Gómez-Niño, A. (2014). Using the Delphi Technique to Improve Science Education in Spain. In C. Bolte, J. Holbrook, R. Mamlok-Naaman, & F. Rauch (Eds.), Science Teachers' Continuous Professional Development in Europe. Case Studies from the PROFILES Project (pp. 31–37). Klagenfurt: Alpen-Adria-Universität Klagenfurt.

Christidou, V. (2011). Interest, Attitudes and Images Related to Science: Combining Students' Voices with the Voices of School Science, Teachers, and Popular Science. International Journal of Environmental and Science Education, 6(2), 141–159.

Coll, R. K., & Taylor, N. (2009). Exploring International Perspectives of Scientific Literacy: An Overview of the Special Issue. International Journal of Environmental & Science Education, 4(3), 197–200.

Cook-Sather, A. (2002). Authorizing Students' Perspectives: Toward Trust, Dialogue, and Change in Education. Educational Researcher, 31(4), 3–14.

Council of Ministers of Education, Canada. (1997). Common Framework of Science Learning Outcomes. Retrieved from http://publications.cmec.ca/science/framework/index.htm [12.11.2013]

Dalkey, N. C. (1969). The Delphi Method I. An Experimental Study of Group Opinion. Santa Monica: RAND Corporation. Retrieved from http://www.rand.org/pubs/research_memoranda/RM5888.html [12.11.2013]

Dalkey, N. C., Brown, B. B., & Cochran, S. W. (1969). The Delphi Method III: Use of Self Ratings to Improve Group Estimates. Santa Monica: RAND Corporation. Retrieved from http://www.rand.org/pubs/research_memoranda/RM6115.html [12.11.2013]

Dalkey, N. C., & Helmer, O. (1963). An Experimental Application of the Delphi Method to the Use of Experts. Management Science, 9(3), 458–467.

DeBoer, G. E. (1991). A History of Ideas in Science Education: Implications for Practice. New York and London: Teachers College Press.

DeBoer, G. E. (1997). Historical Perspectives on Scientific Literacy. In W. Gräber & C. Bolte (Eds.), Scientific Literacy (pp. 69–86). Kiel: IPN.

DeBoer, G. E. (2000). Scientific Literacy. Another Look at its Historical and Contemporary Meanings and its Relationship to Science Education Reform. Journal of Research in Science Teaching, 37(6), 582–601.

Deng, Z. (2007). Scientific Literacy as an Issue of Curriculum Inquiry. In C. Linder, L. Östman, & P.-O. Wickmann, Promoting Scientific Literacy: Science Education Research in Transaction Proceedings of the Linnaeus Tercentenary Symposium held at Uppsala University, Uppsala, Sweden, May 28-29, 2007 (pp. 134–139). Uppsala.

Deutsches PISA-Konsortium. (2001). PISA 2000. Basiskompetenzen von Schülerinnen und Schülern im internationalen Vergleich. Opladen: Leske und Budrich.

Deutsch, M., & Gerard, H. B. (1955). A study of normative and informational social influences upon individual judgment. The Journal of Abnormal and Social Psychology, 51(3), 629–636.

Dewey, J. (1916). Democracy and education: an introduction to the philosophy of education. New York: Macmillan.

Dillon, J. (2009). On scientific literacy and curriculum reform. International Journal of Environmental & Science Education, 4(3), 201–213.

Drechsel, B., Carstensen, C., & Prenzel, M. (2011). The Role of Content and Context in PISA Interest Scales: A study of the embedded interest items in the PISA 2006 science assessment. International Journal of Science Education, 33(1), 73–95.

Duit, R., & Mikelskis-Seifert, S. (Eds.). (2010). Physik im Kontext. Seelze: Friedrich Verlag.

Duranti, A., & Goodwin, C. (1992). Rethinking Context: Language as an Interactive Phenomenon. Cambridge: Cambridge University Press.

EC. (1995). White paper on education and training: Teaching and learning - Towards the learning society. Luxembourg: Office for Official Publications in European Countries.

EC (Ed.). (2004). Europe needs more scientists. Report by the High Level Group on Increasing Human Resources for Science and Technology in Europe. Luxembourg: Office for Official Publications of the European Communities. Retrieved from http://ec.europa.eu/research/conferences/2004/sciprof/pdf/final_en.pdf [10.08.2012]

EC. (2005). Europeans, Science and Technology. Retrieved from http://ec.europa.eu/public_opinion/archives/ebs/ebs_224_report_en.pdf [15.12.2013]

EC. (2007). Science Education Now: A Renewed Pedagogy for the Future of Europe. Brussels. Retrieved from http://ec.europa.eu/research/science-society/document_library/pdf_06/report-rocard-on-science-education_en.pdf[07.12.2012]

EC. (2011). Science Education in Europe: National Policies, Practices and Research. Brussels: EACEA. Retrieved from http://eacea.ec.europa.eu/education/eurydice/documents/thematic_reports/133en.pdf [13.11.2013]

Eckebrecht, D., & Schneeweiß, H. (2003). Naturwissenschaftliche Bildung: Gedanken und Beispiele zur Umsetzung von scientific literacy . Stuttgart: Klett.

Edgren, G. (2006). Developing a competence-based core curriculum in biomedical laboratory science: a Delphi study. Medical Teacher, 28(5), 409–417.

Elmose, S., & Roth, W.-M. (2005). Allgemeinbildung: readiness for living in risk society. Journal of Curriculum Studies, 37(1), 11–34.

Elster, D. (2007). Interessante und weniger interessante Kontexte für das Lernen von Naturwissenschaften. Erste Ergebnisse der deutschen ROSE-Erhebung. Der Mathematische Und Naturwissenschaftliche Unterricht, 60(4), 243–249.

Farmer, E. I. (1995). A Delphi Study of Tech Prep Initiatives in Higher Education: Research Priorities in Teacher Education.

Fensham, P. J. (1985). Science for all: A reflective essay. Journal of Curriculum Studies, 17(4), 415–435.

Fensham, P. J. (2002). Time to change drivers for scientific literacy. Canadian Journal of Science, Mathematics and Technology Education, 2(1), 9–24.

Fensham, P. J. (2007). Competences, from within and without: new challenges and possibilities for scientific literacy. In C. Linder, L. Östman, & P.-O. Wickmann, Promoting Scientific Literacy: Science Education Research in Transaction Proceedings of the Linnaeus Tercentenary Symposium held at Uppsala University, Uppsala, Sweden, May 28-29, 2007 (pp. 113–119). Uppsala.

Fensham, P. J. (2009). Real World Contexts in PISA Science: Implications for Context-Based Science Education. Journal of Research in Science Teaching, 46(8), 884–896.

Festinger, L. (1962). A theory of cognitive dissonance. Stanford: Stanford University Press.

Fielding, M. (Ed.). (2001). Special Issue: Student Voice. Forum, 43(2).

Fielding, M. (2004). Transformative Approaches to Student Voice: Theoretical Underpinnings, Recalcitrant Realities. British Educational Research Journal, 30(2), 295–311.

Fleiss, J. L. (1981). Statistical methods for rates and proportions (2. ed.). New York: Wiley.

Fleming, R. (1989). Literacy for a Technological Age. Science Education, 73(4), 391–404.

Frey, K. (1970). Kriteriensysteme in der Curriculumkonstrkution: begriffliche Grundlagen. In K. Frey (Ed.), Kriterien in der Curriculumkonstruktion. Weinheim: Beltz.

Frey, K. (1974). Integriertes Curriculum Naturwissenschaft der Sekundarstufe I: Projekte und Innovationsstrategien - eine Einführung in die Themenstellung. In K. Frey & K. Blänsdorf, Integriertes Curriculum Naturwissenschaft der Sekundarstufe I: Projekte und Innovationsstrategien. Bericht über das 5. IPN-Symposion (pp. 15–26). Weinheim: Beltz.

Frey, K. (1980a). Das curriculare Delphi-Verfahren. In P. Häußler, K. Frey, L. Hoffmann, J. Rost, & H. Spada, Physikalische Bildung: Eine curriculare Delphi-Studie. Teil I: Verfahren und Ergebnisse. IPN-Arbeitsbericht 41 (pp. 30–34). Kiel: IPN.

Frey, K. (1980b). Das curriculare Konzept. In P. Häußler, K. Frey, L. Hoffmann, J. Rost, & H. Spada, Physikalische Bildung: Eine curriculare Delphi-Studie. Teil I: Verfahren und Ergebnisse. IPN-Arbeitsbericht 41 (pp. 21–29). Kiel: IPN.

Frey, K. (1980c). Kriterien für die Auswahl der Teilnehmer. In P. Häußler, K. Frey, L. Hoffmann, J. Rost, & H. Spada, Physikalische Bildung: Eine curriculare Delphi-Studie. Teil I: Verfahren und Ergebnisse. IPN-Arbeitsbericht 41 (pp. 35–44). Kiel: IPN.

Frey, K. (1989). Integrated Science Education: 20 years on. International Journal of Science Education, 11(1), 3–17.

Gauckler, M. (2014a). Aspects of Science Education from a European Perspective – A meta-analysis of the first and second round of the International PROFILES Curricular Delphi Study on Science Education. Bachelor's thesis.

Gauckler, M. (2014b). Aspects of Science Education from a European Perspective – A meta-analysis of the International PROFILES Delphi Study on Science Education Round 3. Master's thesis.

Gauckler, M., Bolte, C., & Schulte, T. (in press). Aspekte naturwissenschaftlicher Bildung im europäischen Meinungsbild. In S. Bernholt (Ed.), Heterogenität und Diversität -

Vielfalt der Voraussetzungen im naturwissenschaftlichen Unterricht. Zur Didaktik der Physik und Chemie. Probleme und Perspektiven. Kiel: IPN.

Gauckler, M., Schulte, T., & Bolte, C. (2014). Aspects of Science Education from a European Perspective – First Results from a Meta-analysis of the International PRO-FILES Curricular Delphi Study on Science Education. In C. Bolte & F. Rauch (Eds.), Enhancing Inquiry-based Science Education and Teachers' Continuous Professional Development in Europe: Insights and Reflections on the PROFILES Project and other Projects funded by the European Commission (pp. 127–135). Klagenfurt: Alpen-Adria-Universität Klagenfurt.

GDCh (Ed.). (2005). Stärkung der naturwissenschaftlichen Bildung: Empfehlungen der Fachgruppe Chemieunterricht der GDCh für einen durchgängigen naturwissenschaftlichen Unterricht von der Grundschule bis zum Fachunterricht der weiterführenden Schulen.

Gilbert, J. K. (2006). On the Nature of "Context" in Chemical Education. International Journal of Science Education, 28(9), 957–976.

Glaser, B. G., & Strauss, A. L. (1967). The Discovery of Grounded Theory: Strategies for Qualitative Research. Chicago: Aldine.

Gläser, J., & Grit, L. (2010). Experteninterviews und qualitative Inhaltsanalyse. Wiesbaden: Springer VS.

Glasersfeld, E. v. (1993). Das Radikale in Piagets Konstruktivismus. In R. Duit & W. Gräber (Eds.), Kognitive Entwicklung und Lernen der Naturwissenschaften (pp. 46–54). Kiel: IPN.

Glasersfeld, E. v. (1996). Radikaler Konstruktivismus: Ideen, Ergebnisse, Probleme. Frankfurt am Main: Suhrkamp.

Gräber, W. (1995). Anregungen aus der Interessenforschung für den Chemie unter richt - Die Bedeutung des Sachinteresses als Determinante des Interesses am Unterrichtsfach Chemie. In R. Arbinger & R. Jäger (Eds.), Zukunftsperspektiven empirischpädagogischer Forschung. Landau: Empirische Pädagogik.

Gräber, W. (2002). „Scientific Literacy" - Naturwissenschaftliche Bildung in der Diskussion. In P. Döbrich (Ed.), Qualitätsentwicklung im naturwissenschaftlichen Unterricht (pp. 1–28). Frankfurt am Main: Gesellschaft zur Förderung Pädagogischer Forschung.

Gräber, W., & Bolte, C. (1997). Scientific Literacy. An International Symposium. Kiel: IPN.

Gräber, W., & Nentwig, P. (2002). Scientific Literacy - Naturwissenschaftliche Grundbildung in der Diskussion. In W. Gräber, P. Nentwig, T. Koballa, & R. Evans (Eds.), Scientific Literacy. Der Beitrag der Naturwissenschaften zur allgemeinen Bildung (pp. 7–20). Opladen: Leske und Budrich.

Gräber, W., Nentwig, P., Koballa, T., & Evans, R. (Eds.). (2002). Scientific literacy. Der Beitrag der Naturwissenschaften zur allgemeinen Bildung. Opladen: Leske und Budrich.

Greve, W., & Wentura, D. (1997). Wissenschaftliche Beobachtung: eine Einführung. Weinheim: Beltz, Psychologie-Verl.-Union.

Gundem, B. B. (1995). Historical Roots and Contemporary Foundations. In S. Hopmann & K. Riquarts (Eds.), Didaktik and/or Curriculum (pp. 43–56). Kiel: IPN.

Häder, M. (2009). Delphi-Befragungen: Ein Arbeitsbuch. Wiesbaden: Verlag für Sozialwissenschaften.

Häder, M., & Häder, S. (2000). Die Delphi-Methode als Gegenstand methodischer Forschung. In M. Häder & S. Häder (Eds.), Die Delphi-Technik in den Sozialwissenschaften: methodische For-schungen und innovative Anwendungen (pp. 11–31). Wiesbaden: Westdeutscher Verlag.

Hattie, J. (2009). Visible Learning. London: Routledge.

Hattie, J. (2012). Visible learning for teachers: maximizing impact on learning. London: Routledge.

Häußler, P. (1973). Bisherige Ansätze zu disziplinübergreifenden naturwissenschaftlichen Curricula - eine Übersicht. In K. Frey & P. Häußler (Eds.), Integriertes Curriculum Naturwissenschaft: theoretische Grundlagen und Ansätze (pp. 31–69). Weinheim: Beltz.

Häußler, P. (1992). Physikalische Bildung als Menschenbildung: Wunsch und Wirklichkeit. In Physikunterricht und Menschenbildung. Kiel: Institut für die Pädagogik der Naturwissenschaften.

Häußler, P., Frey, K., Hoffmann, L., Rost, J., & Spada, H. (1980). Physikalische Bildung: Eine curriculare Delphi-Studie. Teil I: Verfahren und Ergebnisse. IPN-Arbeitsbericht 41. Kiel: IPN.

Häußler, P., & Hoffmann, L. (2000). A curricular frame for physics education: Development, comparison with students' interests, and impact on students' achievement and self-concept. Science Education, 84(6), 689–705. http://doi.org/10.1002/1098-237X(200011)84:6<689::AID-SCE1>3.0.CO;2-L

Häußler, P., & Rost, J. (1980a). Aussagen zur physikalischen Bildung: Ergebnisse der ersten Runde. In P. Häußler, K. Frey, L. Hoffmann, J. Rost, & H. Spada, Physikalische Bildung: Eine curriculare Delphi-Studie. Teil I: Verfahren und Ergebnisse. IPN-Arbeitsbericht 41 (pp. 131–166). Kiel: IPN.

Häußler, P., & Rost, J. (1980b). Aussagen zur physikalischen Bildung in Form von Kategorienkombinationen: Ergebnisse der 2. Aufgabe. In P. Häußler, K. Frey, L. Hoffmann, J. Rost, & H. Spada, Physikalische Bildung: Eine curriculare Delphi-Studie. Teil I: Verfahren und Ergebnisse. IPN-Arbeitsbericht 41 (pp. 192–220). Kiel: IPN.

Havighurst, R. J. (1981). Developmental tasks and education. New York: Longman.

Heimlich, J. E., Carlson, S. P., & Storksdieck, M. (2011). Building face, construct, and content validity through use of a modified Delphi: adapting grounded theory to build an environmental field days observation tool. Environmental Education Research, 17(3), 287–305.

Heymann, H. W. (1990). Überlegungen zu einem zeitgemäßen Allgemeinbildungskonzept. In H. W. Heymann & W. van Lück (Eds.), Allgemeinbildung und öffentliche Schule: Klärungsversuche (pp. 21–28). Bielefeld: Institut für Didaktik der Mathematik.

Heymann, H. W., van Lück, W., Meyer, M. A., Schulze, T., & Tenorth, H.-E. (1990). Allgemeinbildung als Aufgabe der öffentlichen Schule. In H. W. Heymann & W.

van Lück (Eds.), Allgemeinbildung und öffentliche Schule: Klärungsversuche (pp. 9–20). Bielefeld: Institut für Didaktik der Mathematik.

Hoffmann, L., & Lehrke, M. (1986). Eine Untersuchung über Schülerinteressen an Physik und Technik. Zeitschrift Für Pädagogik, 32(2), 189–204.

Hoffmann, L., & Rost, J. (1980). Die Kategorien zum Aussagenelement "Situation, Kontext, Motiv." In P. Häußler, K. Frey, L. Hoffmann, J. Rost, & H. Spada, Physikalische Bildung: Eine curriculare Delphi-Studie. Teil I: Verfahren und Ergebnisse. IPN-Arbeitsbericht 41 (pp. 65–82). Kiel: IPN.

Hollanders, H., & Soete, L. (2010). UNESCO Science Report 2010. Bonn: Deutsche UNESCO-Kommission.

Hopmann, S., & Riquarts, K. (Eds.). (1995a). Didaktik and/or Curriculum. Kiel: IPN.

Hopmann, S., & Riquarts, K. (1995b). Didaktik and/or Curriculum: Basic Problems of Comparative Didaktik. In S. Hopmann & K. Riquarts (Eds.), Didaktik and/or Curriculum (pp. 9–40). Kiel: IPN.

Hurd, P. D. (1998). Scientific Literacy: New Minds for a Changing World. Science Education, 82(3), 407–416.

Jenkins, E. W. (1999). School science, citizenship and the public understanding of science. International Journal of Science Education, 21(7), 703–710.

Jenkins, E. W. (2005). The Student Voice in Science Education: Research and Issues. Journal of Baltic Science Education, 4(1), 22–30.

Jenkins, E. W. (2006). The Student Voice and School Science Education. Studies in Science Education, 42, 49–88.

Judd, R. C. (1972). Use of Delphi Methods in Higher Education. Technological Forecasting and Social Change, 4, 173–186.

Kapanadze, M., & Slovinsky, E. (2014). Stakeholders' Views on Science Education in Georgia – Curricular Delphi Study. In C. Bolte, J. Holbrook, R. Mamlok-Naaman, & F. Rauch (Eds.), Science Teachers' Continuous Professional Development in Europe. Case Studies from the PROFILES Project (pp. 24–30). Klagenfurt: Alpen-Adria-Universität Klagenfurt.

Keinonen, T., Kukkonen, J., Schulte, T., & Bolte, C. (2014). Stakeholders' Views of Science Education: Finnish PROFILES Curricular Delphi Study – Second Round. In C. P. Constantinou, N. Papadouris, & A. Hadjigeorgiou (Eds.), E-Book Proceedings of the ESERA 2013 Conference: Science Education Research For Evidence-based Teaching and Coherence in Learning. Part 8 (co-ed. M. Ossevoort & J. A. Nielsen) (pp. 160–171). Nicosia, Cyprus: European Science Education Research Association. Retrieved from http://www.esera.org/media/esera2013/Tuula_Keinonen_21 Dec2013.pdf [12.06.2014]

Kenis, D. (1995). Improving group decisions: designing and testing techniques for group decision support systems applying Delphi principles. Doctoral dissertation.

King, D. (2012). New perspectives on context-based chemistry education: using a dialectical sociocultural approach to view teaching and learning. Studies in Science Education, 48(1), 51–87.

Klafki, W. (1964). Das pädagogische Problem des Elementaren und die Theorie der kategorialen Bildung (4th ed.). Weinheim: Beltz.

Klafki, W. (1995a). Didactic analysis as the core of preparation of instruction (Didaktische Analyse als Kern der Unterrichtsvorbereitung). Journal of Curriculum Studies, 27(1), 13–30.

Klafki, W. (1995b). On the Problem of Teaching and Learning Contents from the Standpoint of Critical-Constructive Didaktik. In S. Hopmann & K. Riquarts (Eds.), Didaktik and/or Curriculum (pp. 187–200). Kiel: Institut für die Pädagogik der Naturwissenschaften.

Klafki, W. (2000). The significance of classical theories of Bildung for a contemporary concept of Allgemeinbildung. In I. Westbury, S. Hopmann, & K. Riquarts (Eds.), Teaching as a Reflective Practice: The German Didaktik Tradition (pp. 85–108). Mahwah, NJ: Lawrence Erlbaum Associates.

Klafki, W. (2007). Neue Studien zur Bildungstheorie und Didaktik: zeitgemäße Allgemeinbildung und kritisch-konstruktive Didaktik (6. ed.). Weinheim: Beltz.

Klemm, K., Rolff, H.-G., & Tillmann, K.-J. (1985). Bildung für das Jahr 2000. Bilanz der Reform, Zukunft der Schule. Reinbek: Rowohlt.

Klieme, E., Avenarius, H., Blum, W., Döbrich, P., Gruber, H., Prenzel, M., ... Vollmer, H. J. (2007). Zur Entwicklung nationaler Bildungsstandards. (Bundesministerium für Bildung und Forschung, Ed.). Bonn.

KMK. (2005a). Bildungsstandards im Fach Biologie für den Mittleren Schulabschluss. München, Neuwied: Luchterhand. Retrieved from http://www.kmk.org/fileadmin/veroeffentlichungen_beschluesse/2004/2004_12_16-Bildungsstandards-Biologie.pdf [30.07.2011]

KMK. (2005b). Bildungsstandards im Fach Chemie für den Mittleren Schulabschluss. München, Neuwied: Luchterhand. Retrieved from http://www.kmk.org/fileadmin/veroeffentlichungen_beschluesse/2004/2004_12_16-Bildungsstandards-Chemie.pdf [30.07.2011]

KMK. (2005c). Bildungsstandards im Fach Physik für den Mittleren Schulabschluss. München, Neuwied: Luchterhand. Retrieved from http://www.kmk.org/fileadmin/veroeffentlichungen_beschluesse/2004/2004_12_16-Bildungsstandards-Mittleren-SA-Bio-Che-Phy.pdf [30.07.2011]

Koballa, T., Kemp, A., & Evans, R. (1997). The Spectrum of Scientific Literacy. Science Teacher, 64(7), 27–31.

Kolstoe, S. D. (2000). Consensus projects: teaching science for citizenship. International Journal of Science Education, 22(6), 645–664.

Kolstø, S. D. (2001). Scientific literacy for citizenship: Tools for dealing with the science dimension of controversial socioscientific issues. Science Education, 85(3), 291–310.

Kortland, J. (2011). Scientific Literacy and Context-Based Curricula: Exploring the Didactical Friction between Context and Science Knowledge. In D. Höttecke, Naturwissenschaftliche Bildung als Beitrag zur Gestaltung partizipativer Demokratie (pp. 17–31). Berlin: LIT Verlag.

Kremer, A., & Stäudel, L. (1997). Zum Stand des fächerübergreifendenden naturwissenschaftlichen Unterrichts in der Bundesrepublik Deutschland. Eine vorläufige Bilanz. Zeitschrift Für Didaktik Der Naturwissenschaften, 3(3), 52–66.

Kremer, M. (2012). Grundbildung in den naturwissenschaftlichen Fächern – Basiskompetenzen. MNU. Retrieved from http://www.mnu.de/mnu-publikationen/publikationen [25.01.2014]

Krüger, D., & Riemeier, T. (2014). Die qualitative Inhaltsanalyse - eine Methode zur Auswertung von Interviews. In D. Krüger, I. Parchmann, & H. Schecker (Eds.), Methoden in der naturwissenschaftsdidaktischen Forschung (pp. 133–145). Berlin & Heidelberg: Springer.

Labovitz, S. (1967). Some Observations on Measurement and Statistics. Social Forces, 46(2), 151–160. http://doi.org/10.2307/2574595

Labudde, P., Heitzmann, A., Heiniger, P., & Widmer, I. (2005). Dimensionen und Facetten des fächerübergreifenden naturwissenschaftlichen Unterrichts: ein Modell. Zeitschrift für Didaktik der Naturwissenschaften, 11, 103–115.

Labudde, P., & Möller, K. (2012). Stichwort: Naturwissenschaftlicher Unterricht. Zeitschrift Für Erziehungswissenschaft, 1, 11–36.

Landis, J. R., & Koch, G. G. (1977). The Measurement of Observer Agreement for Categorical Data. Biometrics, 33(1), 159–174.

Laugksch, R. C. (2000). Scientific literacy: A conceptual overview. Science Education, 84(1), 71–94.

Lauterbach, R. (1992a). Physikunterricht: Von der Qualifizierung zur Bildung? In P. Häußler (Ed.), Physikunterricht und Menschenbildung (pp. 13–36). Kiel: IPN.

Lauterbach, R. (1992b). Praxis integrierter naturwissenschaftlicher Grundbildung (PING). In P. Häußler (Ed.), Physikunterricht und Menschenbildung (pp. 251–268). Kiel: Institut für die Pädagogik der Naturwissenschaften.

Lauterbach, R. (1993). Konzepte für eine naturwissenschaftlich-technische Grundbildung. Habilita-tionsschrift.

Linstone, H. A., & Turoff, M. (1975a). Introduction to chapter I: Introduction. In H. A. Linstone & M. Turoff (Eds.), The Delphi Method: Techniques and Applications (pp. 3–12). Reading, Massachusetts: Addison-Wesley.

Linstone, H. A., & Turoff, M. (1975b). Introduction to chapter IV: Evaluation. In H. A. Linstone & M. Turoff (Eds.), The Delphi Method: Techniques and Applications (pp. 229–236). Reading, Massachusetts: Addison-Wesley.

Linstone, H. A., & Turoff, M. (Eds.). (1975c). The Delphi Method: Techniques and Applications. Reading, Massachusetts: Addison-Wesley.

Litt, T. (1959). Naturwissenschaft und Menschenbildung (3. ed.). Heidelberg: Quelle und Meyer.

Løvlie, L., & Standish, P. (2002). Introduction: Bildung and the idea of a liberal education. Journal of Philosophy of Education, 36(3), 317–340.

Marshall, A. P., Currey, J., Aitken, L. M., & Elliott, D. (2007). Key stakeholders' expectations of educational outcomes from Australian critical care nursing courses: A Delphi study. Australian Critical Care, 20(3), 89–99.

Martin, M. O., Mullis, I. V. S., Foy, P., & Stanco, G. M. (2012). TIMSS 2011 International Results in Science. Chestnut Hill, MA: TIMSS & PIRLS International Study Center, Boston College. Retrieved from http://timss.bc.edu/timss2011/downloads/ T11_IR_Science_FullBook.pdf [08.12.2013]

Mayer, H. O. (2006). Interview und schriftliche Befragung: Entwicklung, Durchführung und Auswertung. München: Oldenbourg.

Mayer, J. (1992). Formenvielfalt im Biologieunterricht: Ein Vorschlag zur Neubewertung der Formenkunde. Kiel: IPN.

Mayring, P. (1983). Qualitative Inhaltsanalyse: Grundlagen und Techniken. Weinheim: Beltz.

Mayring, P., & Gläser-Zikuda, M. (Eds.). (2008). Die Praxis der Qualitativen Inhaltsanalyse. Weinheim: Beltz.

Meyer, M. A. (2005). Die Bildungsgangforschung als Rahmen für die Weiterentwicklung der allgemeinen Didaktik. In B. Schenk (Ed.), Bausteine einer Bildungsgangtheorie (pp. 17–46). VS Verlag für Sozialwissenschaften. Retrieved from http://link.springer.com/chapter/10.1007/978-3-322-80754-0_2

Meyer, M. A., & Reinartz, A. (1998). Bildungsgangdidaktik. Denkanstöße für pädagogische Forschung und schulische Praxis. Opladen: Leske und Budrich.

Millar, R. (1996). Towards a Science Curriculum for Public Understanding. School Science Review, 77(280), 7–18.

Millar, R. (2005). Contextualized science courses: Where next? In P. Nentwig & D. J. Waddington (Eds.), Making it relevant. Context-based learning of science (pp. 323–346). Münster: Waxmann.

Millar, R. (2006). Twenty First Century Science: Insights from the Design and Implementation of a Scientific Literacy Approach in School Science. International Journal of Science Education, 28(13), 1499–1521.

Millar, R., & Osborne, J. F. (Eds.). (1998). Beyond 2000: Science Education for the Future. London: King's College London, School of Education. Retrieved from http://www.nationalstemcentre.org.uk/elibrary/resource/6929/beyond-2000-science-education-for-the-future [18.07.2014]

Miller, G. A. (1956). The Magical Number Seven, Plus or Minus Two: Some Limits on Our Capacity for Processing Information. Psychological Review, 63(2), 81–97.

MNU (Ed.). (2003). Lernen und Können im naturwissenschaftlichen Unterricht. Denkanstöße und Empfehlungen zur Entwicklung von Bildungs-Standards in den naturwissenschaftlichen Fächern Biologie, Chemie und Physik (Sekundarbereich I). Retrieved from https://www.mnu.de/images/Dokumente/rubberdoc/mnupublbildstand natwiss.pdf [25.01.2014]

Mogensen, F., & Schnack, K. (2010). The action competence approach and the "new" discourses of education for sustainable development, competence and quality criteria. Environmental Education Research, 16(1), 59–74.

Murry, J. W., & Hammons, J. O. (1995). Delphi: A Versatile Methodology for Conducting Qualitative Research. Review of Higher Education, 18(4), 423–36.

National Science Teachers Association. (1991). Position Statement. Washington DC: National Science Teachers Association.

Neuner, G. (1999). Ressource Allgemeinbildung? Neue Aktualität eines alten Themas. Weinheim: Deutscher Studien Verlag.

Nijhof, W. J. (1990). Values in higher education. "Bildungsideale" in historical and contemporary perspective. Enschede: University of Twente, Department of Education.

Nixon, J., Martin, J., McKeon, P., & Ranson, S. (1996). Encouraging Learning: Towards a Theory of the Learning School. Buckingham: Open University Press.

Norris, S. P., & Phillips, L. M. (2003). How literacy in its fundamental sense is central to scientific literacy. Science Education, 87(2), 224–240.

NRC. (1996). National Science Education Standards. Washington, DC: National Academy of Sciences.

NRC. (2000). Inquiry and the National Science Education Standards: A Guide for Teaching and Learning. Washington DC: National Academy of Sciences.

Nworie, J. (2011). Using the Delphi Technique in Educational Technology Research. TechTrends, 55(5), 24–30.

OECD. (1999). Measuring Student Knowledge and Skills. Paris: OECD. Retrieved from http://www.oecd.org/edu/school/programmeforinternationalstudentassessmentpisa/3 3693997.pdf [11.10.2013]

OECD. (2000). Measuring Student Knowledge and Skills. Paris: OECD. Retrieved from http://www.oecd-ilibrary.org/content/book/9789264181564-en [11.10.2013]

OECD. (2001). Definition and Selection of Competencies: Theoretical and Conceptual Foundations (DeDeCo). Background Paper. Retrieved from http://www.oecd.org/education/skills-beyond-school/41529556.pdf [12.10.2014]

OECD. (2002). Definition and Selection of Competences (DeSeCo). Strategy Paper. Retrieved from http://www.deseco.admin.ch/bfs/deseco/en/index/02.parsys. 34116.downloadList.87902.DownloadFile.tmp/oecddesecostrategypaperdeelsaedceri cd20029.pdf [12.10.2014]

OECD. (2003). Literacy Skills for the World of Tomorrow: Further Results from PISA 2000. Paris: OECD. Retrieved from http://www.oecd.org/education/ preschoolandschool/programmeforinternationalstudentassessmentpisa/33690591.pdf [01.10.2012]

OECD. (2004a). Learning for Tomorrow's World. First Results from PISA 2003. Paris: OECD. Retrieved from http://www.oecd.org/dataoecd/1/60/34002216.pdf [01.10.2012]

OECD. (2004b). The PISA 2003 Assessment Framework. Paris: OECD. Retrieved from http://www.oecd-ilibrary.org/content/book/9789264101739-en [11.10.2013]

OECD. (2005). The Definition and Selection of Key Competencies. Executive Summary. Retrieved from http://www.oecd.org/pisa/35070367.pdf [12.10.2014]

OECD. (2006). Assessing Scientific, Reading and Mathematical Literacy. A Framework for PISA 2006. Paris: OECD. Retrieved from http://www.oecd-ilibrary.org/content/book/9789264026407-en [11.10.2013]

OECD. (2007a). PISA 2006. Naturwissenschaftliche Kompetenzen für die Welt von Morgen. Paris: OECD. Retrieved from http://www.oecd.org/pisa/39731064.pdf [01.10.2012]

OECD. (2007b). PISA 2006: Science Competencies for Tomorrow's World. Paris: OECD. Retrieved from http://www.oecd-ilibrary.org/content/book/9789264040014-en [11.10.2013]

OECD. (2010). PISA 2009 Results: What Students Know and Can Do – Student Performance in Reading, Mathematics and Science (Volume I). Paris: OECD. Retrieved from http://www.oecd.org/pisa/pisaproducts/48852548.pdf [01.10.2012]

OECD. (2013). PISA 2015 Draft Science Framework. Retrieved from http://www.oecd.org/callsfortenders/Annex%20IA_%20PISA%202015%20Science%20Framework%20.pdf [24.09.2014]

OECD. (2014). PISA 2012 Results in Focus. What 15-year-olds know and what they can do with what they know. Paris: OECD. Retrieved from http://www.oecd.org/pisa/keyfindings/pisa-2012-results.htm [01.03.2015]

Okoli, C., & Pawlowski, S. D. (2004). The Delphi method as a research tool: an example, design considerations and applications. Information & Management, 42(1), 15–29.

Oliver, J. S., Jackson, D. F., & Chun, S. (2001). The Concept of Scientific Literacy: A View of the Current Debate as an Outgrowth of the Past Two Centuries. Electronic Journal of Literacy Through Science, 1(1), 1–33. Retrieved from http://ejlts.ucdavis.edu/article/2001/1/1/concept–scientific–literacy–view–current–debate–outgrowth–past–two–centuries [10.11.2013].

Osborne, J. F. (2007). Science Education for the Twenty First Century. Eurasia Journal of Mathematics, Science & Technology Education, 3(3), 173–184.

Osborne, J. F., & Collins, S. (2000). Pupils' and Parents' Views of the School Science Curriculum. London: King's College London.

Osborne, J. F., & Collins, S. (2001). Pupils' views of the role and value of the science curriculum: A focus-group study. International Journal of Science Education, 23(5), 441–467.

Osborne, J. F., Ratcliffe, M., Collins, S., Millar, R., & Duschl, R. (2003). What "'Ideas-about-Science'" Should Be Taught in School Science? A Delphi Study of the Expert Community. Journal of Research in Science Teaching, 40(7), 692–720.

Ozdem, Y., & Cavas, B. (2014). The Realization of Inquiry-based Science Education in PROFILES: Using a Delphi study to Guide Continuous Professional Development. In C. Bolte, J. Holbrook, R. Mamlok-Naaman, & F. Rauch (Eds.), Science Teachers' Continuous Professional Development in Europe. Case Studies from the PROFILES Project (pp. 59–67). Klagenfurt: Alpen-Adria-Universität Klagenfurt.

Parchmann, I., Gräsel, C., Baer, A., Nentwig, P., Demuth, R., & Ralle, B. (2006). "Chemie im Kontext": A symbiotic implementation of a context-based teaching and learning approach. International Journal of Science Education, 28(9), 1041–1062.

Parenté, F. J., & Anderson-Parenté, J. K. (1987). Delphi Inquiry Systems. In G. Wright & P. Ayton (Eds.), Judgmental Forecasting (pp. 129–157). New York: John Wiley & Sons.

Phillips, D. C., & Siegel, H. (2013). Philosophy of Education. Retrieved October 9, 2014, from http://plato.stanford.edu/archives/win2013/entries/education-philosophy [02.10.2014]

Pinar, W. F. (2009). Bildung and the Internationalization of Curriculum Studies. In E. Ropo & T. Autio, International Conversations on Curriculum Studies. Subject, Society and Curriculum (pp. 23–41). Rotterdam et al.: Sense Publishers.

Posner, G. (1995). Curriculum Theory, School Science and the Natural Sciences. In S. Hopmann & K. Riquarts (Eds.), Didaktik and/or Curriculum (pp. 345–357). Kiel: IPN.

Prenzel, M. (1988). Die Wirkungsweise von Interesse: ein pädagogisch-psychologisches Erklärungsmodell. Opladen: Westdeutscher Verlag.

Prenzel, M. (2010). Naturwissenschaftlicher Fachunterricht. In G. Schaefer (Ed.), Allgemeinbildung durch Naturwissenschaften: Denkschrift der GDNÄ-Bildungskommission (pp. 21–24). Köln: Aulis-Verlag Deubner.

PROFILES. (2010a). FP7 Negotiation Guidance Notes - Coordination and Support Actions. Annex I - Description of Work. Unpublished.

PROFILES. (2010b). The PROFILES Project. Retrieved from http://www.profiles-project.eu [11.06.2012]

Ramsey, J. M. (1997). STS Issue Instruction: Meeting the Goal of Social Responsibility in a Context of Scientific Literacy. In W. Gräber & C. Bolte (Eds.), Scientific Literacy (pp. 305–330). Kiel: Institut für die Pädagogik der Naturwissenschaften.

Rasch, B., Friese, M., Hofmann, W., & Naumann, E. (2006). Quantitative Methoden 2: Einführung in die Statistik (2. ed.). Berlin, Heidelberg: Springer.

Ratcliffe, M., & Millar, R. (2009). Teaching for understanding of science in context: Evidence from the pilot trials of the Twenty First Century Science courses. Journal of Research in Science Teaching, 46(8), 945–959.

Reeves, G., & Jauch, D. L. R. (1978). Curriculum development through Delphi. Research in Higher Education, 8(2), 157–168.

Rice, K. (2009). Priorities in K-12 Distance Education: A Delphi Study Examining Multiple Perspectives on Policy, Practice, and Research. Educational Technology & Society, 12(3), 163–177.

Riquarts, K., Dierks, W., Duit, R., Eulefeld, G., Haft, H., & Stork, H. (Eds.). (1994). Naturwissenschaftliche Bildung in der Bundesrepublik Deutschland. Band II: Naturwissenschaftliche Bildung in öffentlichen und privaten Institutionen. Kiel: IPN.

Riquarts, K., & Wadewitz, C. (2001). Framework for science education in Germany (3., rev. ed). Kiel: IPN.

Roberts, D. A. (1988). What counts as science education? In P. J. Fensham (Ed.), Developments and Dilemmas in Science Education (pp. 27–54). London: Falmer.

Roberts, D. A. (2007). Opening remarks. In C. Linder, L. Östman, & P.-O. Wickmann (Eds.), Promoting scientific literacy: Science education research in transaction. Proceedings of the Linnaeus Tercentenary Symposium (pp. 9–17). Uppsala: Uppsala University.

Robertson, M., Line, M., Jones, S., & Thomas, S. (2000). International Students, Learning Environments and Perceptions: A case study using the Delphi technique. Higher Education Research & Development, 19(1), 89–102.

Robinsohn, S. B. (1975). Bildungsreform als Revision des Curriculum. Neuwied: Luchterhand.

Rockwell, K., Furgason, J., & Marx, D. B. (2000). Research and Evaluation Needs for Distance Education: A Delphi Study. Online Journal of Distance Learning Administration, 3(3).

Rosenberg, M. J., & Hovland, C. I. (1960). Cognitive, Affective and Behavioral Components of Atti-tudes. In M. J. Rosenberg, C. I. Hovland, W. J. McGuire, R. P. Abelson, & J. W. Brehm (Eds.), Attitude Organization and Change: An Analysis of Consistency among Attitude Components. New Haven: Yale University Press.

Rost, J. (2004). Lehrbuch Testtheorie - Testkonstruktion. Bern: Huber.

Rost, J., Senkbeil, M., Walter, O., Carstensen, C. H., & Prenzel, M. (2005). Naturwissenschaftliche Grundbildung im Ländervergleich. In M. Prenzel, J. Baumert, W. Blum, R. Lehmann, D. Leutner, & M. Neubrand (Eds.), PISA 2003. Der zweite Vergleich der Länder in Deutschland – Was wissen und können Jugendliche (pp. 103–124). Münster: Waxmann.

Rost, J., & Spada, H. (1980). Soll- und Ist-Einschätzungen der Aussagenbündel: Ergebnisse der 1. Aufgabe. In P. Häußler, K. Frey, L. Hoffmann, J. Rost, & H. Spada, Physikalische Bildung: Eine curriculare Delphi-Studie. Teil I: Verfahren und Ergebnisse. IPN-Arbeitsbericht 41 (pp. 180–192). Kiel: IPN.

Rowe, G., & Wright, G. (1999). The Delphi technique as a forecasting tool: issues and analysis. International Journal of Forecasting, 15(4), 353–375.

Rowe, G., & Wright, G. (2001). Expert Opinions in Forecasting: The Role of the Delphi Technique. Principles of Forecasting, 30, 125–144.

Ruddock, J. (2003). Consulting Pupils About Teaching and Learning. Teaching and Learning Research Programme. Retrieved from http://www.tlrp.org/pub/documents/no5_ruddock.pdf [13.09.2014]

Ruddock, J., Arnot, M., Fielding, M., MacBath, J., McIntyre, D., Myers, K., … Flutter, J. (2003). Con-sulting Pupils about Teaching and Learning. Projcet homepage. Teaching and Learning Research Programme. Retrieved from http://www.tlrp.org/proj/phase1/phase1dsept.html [13.09.2014]

Rudduck, J., & Fielding, M. (2006). Student voice and the perils of popularity. Educational Review, 58(2), 219–231.

Rundgren, C.-J., Persson, T., & Chang-Rundegren, S.-N. (2014). Comparing Different Stakeholders' View on Science Education with the Science Curriculum in Sweden. In C. Bolte, J. Holbrook, R. Mamlok-Naaman, & F. Rauch (Eds.), Science Teachers' Continuous Professional Development in Europe. Case Studies from the PROFILES Project (pp. 38–47). Klagenfurt: Alpen-Adria-Universität Klagenfurt.

Rychen, D. S. (2008). OECD Referenzrahmen für Schlüsselkompetenzen - ein Überblick. In G. de Haan & I. Bormann (Eds.), Kompetenzen der Bildung für nachhaltige Entwicklung: Operationalisierung, Messung, Rahmenbedingungen, Befunde (pp. 15–22). Wiesbaden: Verlag für Sozialwissenschaften.

Rychen, D. S., & Salganik, L. H. (2002). DeSeCo Symposium - Discussion Paper. Retrieved from http://www.deseco.admin.ch/bfs/deseco/en/index/04.parsys.29226. downloadList.67777.DownloadFile.tmp/2002.desecodiscpaperjan15.pdf#page=5& zoom=100,-78,713 [12.10.2014]

Sadler, T. D. (2004). Informal reasoning regarding socioscientific issues: A critical review of research. Journal of Research in Science Teaching, 41(5), 513–536.

Sadler, T. D., & Zeidler, D. L. (2009). Scientific literacy, PISA, and socioscientific discourse: Assessment for progressive aims of science education. Journal of Research in Science Teaching, 46(8), 909–921.

Salganik, L. H., Rychen, D. S., Moser, U., & Konstant, J. W. (1999). Definition and Selection of Competencies. Projects on Competencies in the OECD Context. Analysis of Theoretical and Conceptual Foundations. Neuchatel: Swiss Federal Statistical Office. Retrieved from http://www.deseco.admin.ch/bfs/deseco/en/index/02.parsys. 53466.downloadList.62701.DownloadFile.tmp/1999.projectsoncompetenciesanalysi s.pdf [12.102014]

Schaefer, G. (2010a). Allgemeinbildung durch Naturwissenschaften - das Konzept eines "fachübergreifenden Fachunterrichts." In G. Schaefer (Ed.), Allgemeinbildung durch Naturwissenschaften: Denkschrift der GDNÄ-Bildungskommission (pp. 9–15). Köln: Aulis-Verlag.

Schaefer, G. (2010b). General Education Through Science Teaching: Memorandum "Allgmeinbildung Durch Naturwissenschaften" (2007) of the Educational Commission of Gesellschaft Deutscher Naturforscher und Ärzte (GDNÄ) (Society of German Natural Researchers and Medicals). Köln: Aulis-Verlag Deubner.

Schaller, K. (1995). The Didactic of J. A. Comenius Between Instruction Technology and Pansophy. In S. Hopmann & K. Riquarts (Eds.), Didaktik and/or Curriculum (pp. 57–69). Kiel: IPN.

Schecker, H., Bethge, T., Breuer, E., Dwingelo-Lütten, R., Graf, H. U., Gropengiesser, I., & Langensiepen, B. (1996). Naturwissenschaftlicher Unterricht im Kontext allgemeiner Bildung. Der Mathematische Und Naturwissenschaftliche Unterricht, 49(8), 488–492.

Schenk, B. (Ed.). (2005). Bausteine einer Bildungsgangdidaktik. Verlag für Sozialwissenschaften

Schenk, B. (2007). Fachkultur und Bildung in den Fächern Chemie und Physik. In J. Lüders (Ed.), Fachkulturforschung in der Schule (pp. 83–100). Barbara Budrich.

Schöler, W. (1970). Geschichte des naturwissenschaftlichen Unterrichts im 17. bis 19. Jahrhundert: Erziehungstheoretische Grundlegung und schulgeschichtliche Entwicklung. Berlin: de Gruyter.

Schreiner, C., & Sjøberg, S. (2004). ROSE. The Relevance of Science Education. Retrieved from http://www.uv.uio.no/ils/english/research/projects/rose/actadidactica. pdf [01.12.2011]

Schulte, T., & Bolte, C. (2012). European Stakeholders Views on Inquiry Based Science Education – Method of and Results from the International PROFILES Curricular Delphi Study on Science Education Round 1. In C. Bolte, J. Holbrook, & F. Rauch (Eds.), Inquiry-based Science Education in Europe - Reflections from the PROFILES Project (pp. 42–51). Berlin: Freie Universität Berlin.

Schulte, T., & Bolte, C. (2013a). Naturwissenschaftliche Bildung im Meinungsbild internationaler Stakeholder. In S. Bernholt (Ed.), Inquiry-based learning - Forschendes Lernen. Gesellschaft für Didaktik der Chemie und Physik. Jahrestagung in Hannover 2012 (pp. 189–191). Kiel: IPN. Retrieved from http://www.gdcp.de/index.php/

tagungsbaende/tagungsband-uebersicht/145-tagungsbaende/2013/4220-band33 [06.03.2013]

Schulte, T., & Bolte, C. (2013b). Views from Different Stakeholders within the "PRO-FILES Curricular Delphi Study on Science Education." In Paper presented at the Annual Meeting of the National Association for the Research on Science Teaching (NARST). Puerto Rico: NARST.

Schulte, T., & Bolte, C. (2014a). Case Studies on Science Education based on Stakeholders' Views Obtained by Means of a National/International PROFILES Curricular Delphi Study. In C. Bolte, J. Holbrook, R. Mamlok-Naaman, & F. Rauch (Eds.), Science Teachers' Continuous Professional Development in Europe. Case Studies from the PROFILES Project (pp. 19–23). Klagenfurt: Alpen-Adria-Universität Klagenfurt.

Schulte, T., & Bolte, C. (2014b). Stakeholders' Views on Empirically based Concepts for Science Education to Enhance Scientific Literacy – Results from the Third Round of the International PROFILES Curricular Delphi Study on Science Education. In C. P. Constantinou, N. Papadouris, & A. Hadjigeorgiou (Eds.), E-Book Proceedings of the ESERA 2013 Conference: Science Education Research For Evidence-based Teaching and Coherence in Learning. Part 8 (co-ed. M. Ossevoort & J. A. Nielsen) (pp. 197–204). Nicosia, Cyprus: European Science Education Research Association. Retrieved from http://www.esera.org/media/esera2013/Theresa_Schulte2_19Dec 2013.pdf [12.06.2014]

Schulte, T., Bolte, C., Keinonen, T., Gorghiu, G., Kapanadze, M., & Charro, E. (2014). A Comparative Analysis of Stakeholders' Views on Science Education from Five Different Partner Countries – Results of the Second Round of the International PRO-FILES Curricular Delphi Study on Science Education. In C. P. Constantinou, N. Papadouris, & A. Hadjigeorgiou (Eds.), E-Book Proceedings of the ESERA 2013 Conference: Science Education Research For Evidence-based Teaching and Coherence in Learning. Part 8 (co-ed. M. Ossevoort & J. A. Nielsen) (pp. 185–196). Nicosia, Cyprus: European Science Education Research Association. Retrieved from http://www.esera.org/media/esera2013/Theresa_Schulte1_10Feb2014.pdf [12.06.2014]

Schulte, T., Georgiu, Y., Kyza, E. A., & Bolte, C. (2014). Students' and Teachers' Perceptions of School-based Scientific Literacy Priorities and Practice: A Cross-Cultural Comparison between Cyprus and Germany. In C. P. Constantinou, N. Papadouris, & A. Hadjigeorgiou (Eds.), E-Book Proceedings of the ESERA 2013 Conference: Science Education Research For Evidence-based Teaching and Coherence in Learning. Part 10 (co-ed. J. Dillon & A. Redfors) (pp. 132–140). Nicosia, Cyprus: European Science Education Research Association. Retrieved from http://www.esera.org/media/esera2013/Theresa_Schulte_19Dec2013.pdf [12.06.2014]

Schulze, T. (1990). Thesen zur Allgemeinbildung. In H. W. Heymann & W. van Lück (Eds.), Allgemeinbildung und öffentliche Schule: Klärungsversuche (pp. 93–110). Bielefeld: Institut für Didaktik der Mathematik.

Seeger, T. (1979). Die Delphi-Methode: Expertenbefragung zwischen Prognose und Gruppenmeinungsbildungsprozesse, überprüft am Beispiel von Delphi-Befragungen im Gegenstandsbereich Information und Dokumentation.

Shamos, M. H. (1995). The myth of scientific literacy. New Brunswick, NJ: Rutgers University Press.

Shamos, M. H. (2002). Durch Prozesse ein Bewusstsein für die Naturwissenschaften entwickeln. In W. Gräber, P. Nentwig, T. Koballa, & R. Evans (Eds.), Scientific Literacy. Der Beitrag der Naturwissenschaften zur allgemeinen Bildung (pp. 45–68). Opladen: Leske und Budrich.

Shen, B. S. P. (1975). Science Literacy: Public understanding of science is becoming vitally needed in developing and industrialized countries alike. American Scientist, 63(3), 265–268.

Sjøberg, S., & Schreiner, C. (2010). The ROSE project. An overview and key findings. Retrieved from http://roseproject.no/network/countries/norway/eng/nor-Sjoberg-Schreiner-overview-2010.pdf [11.01.2014]

Sjöström, J. (2013). Towards Bildung-Oriented Chemistry Education. Science & Education, 22(7), 1873–1890.

Skulmoski, G. J., Hartman, F. T., & Krahn, J. (2007). The Delphi Method for Graduate Research. Journal of Information Technology Education, 6, 1–21.

So, H.-J. S., & Bonk, C. J. (2010). Examining the Roles of Blended Learning Approaches in Computer-Supported Collaborative Learning (CSCL) Environments: A Delphi Study. Educational Technology & Society, 13, 189–200.

Spada, H. (1980). Die Kategorien zum Aussagenelement "Verfügbarkeit." In P. Häußler, K. Frey, L. Hoffmann, J. Rost, & H. Spada, Physikalische Bildung: Eine curriculare Delphi-Studie. Teil I: Verfahren und Ergebnisse. IPN-Arbeitsbericht 41 (pp. 113–130). Kiel: IPN.

Streller, S. (2009). Förderung von Interesse an Naturwissenschaften: eine empirische Untersuchung zur Entwicklung naturwissenschaftlicher Interessen von Grundschulkindern im Rahmen eines außerschulischen Lernangebots. Frankfurt am Main: Lang.

Sühl-Strohmenger, W. (1984). Horizonte von Bildung und Allgemeinbildung: der Bildungsbegriff der Gegenwart im Brennpunkt von Persönlichkeits-, Gesellschafts- und Wissenschaftsorientierung, Konsequenzen für das Verständnis von Allgemeinbildung heute und für die Lehrplangestaltung. Frankfurt am Main: Lang.

Symington, D., & Tytler, R. (2004). Community leaders' views of the purposes of science in the compulsory years of schooling. International Journal of Science Education, 26(11), 1403–1418.

Tenorth, H.-E. (Ed.). (1986a). Allgemeine Bildung: Analysen zu ihrer Wirklichkeit, Versuche über ihre Zukunft. Weinheim; München: Juventa-Verlag.

Tenorth, H.-E. (1986b). Bildung, allgemeine Bildung, Allgemeinbildung. In H.-E. Tenorth (Ed.), Allgemeine Bildung: Analysen zu ihrer Wirklichkeit, Versuche über ihre Zukunft. Weinheim; München: Juventa-Verlag.

Tenorth, H.-E. (1994). "Alle alles zu lehren": Möglichkeiten und Perspektiven allgemeiner Bildung. Darmstadt: Wissenschaftliche Buchgesellschaft.

Tenorth, H.-E. (Ed.). (2003). Form der Bildung - Bildung der Form. Weinheim: Beltz.

Tenorth, H.-E. (2006). Erziehung zur Persönlichkeit. In H.-E. Tenorth, M. Hüther, & M. Heimbach-Steins (Eds.), Erziehung und Bildung heute (pp. 7–24). Berlin: Verlag der GDA.

The RAND Corporation. (2013). Retrieved from www.rand.org [11.10.2013]

TLRP. (2006). Science education in schools. Issues, evidence and proposals. Retrieved from http://www.tlrp.org/pub/documents/TLRP_Science_Commentary_FINAL.pdf [12.08.2014]

Tyler, R. W. (1971). Basic principles of curriculum and instruction. Chicago: Univ. Press.

UNESCO (Ed.). (1993). Project 2000+. International Forum on Scientific and Technological Literacy for All. Paris.

UNESCO. (2012). International Standard Classification of Education. ISCED 2011. Retrieved from http://www.uis.unesco.org/Education/Documents/isced-2011-en.pdf [04.05.2014]

United Nations Environment Programme. (2012). 21 Issues for the 21st Century: Results of the UNEP Foresight Process on Emerging Environmental Issues. Nairobi, Kenya.

van Zolingen, S. J., & Klaassen, C. A. (2003). Selection processes in a Delphi study about key qualifications in Senior Secondary Vocational Education. Technological Forecasting and Social Change, 70(4), 317–340.

Vásquez-Levy, D. (2002). Bildung-centred Didaktik: A framework for examining the educational potential of subject matter. Journal of Curriculum Studies, 34(1), 117–128.

von Engelhardt, D. (2010). Naturwissenschaftliche Bildung - Stationen und Aspekte der Entwicklung. In G. Schaefer (Ed.), Allgemeinbildung durch Naturwissenschaften: Denkschrift der GDNÄ-Bildungskommission (pp. 15–21). Köln: Aulis-Verlag Deubner.

von Hentig, H. (1996). Bildung: ein Essay. München, Wien: Carl Hanser Verlag.

Waddington, D. J. (2005). Context-based learning in science education: a review. In P. Nentwig & D. J. Waddington (Eds.), Making it relevant. Context-based learning of science (pp. 305–334). Münster: Waxmann.

Wagenschein, M. (1968). Verstehen lehren. Weinheim: Beltz.

Wagenschein, M. (1980). Naturphänomene sehen und verstehen: genetische Lehrgänge (1st ed.). Stuttgart: Klett.

Walberg, H. J., & Paik, S. (1997). Scientific Literacy as an International Concern. In W. Gräber & C. Bolte (Eds.), Scientific Literacy (pp. 143–166). Kiel: IPN.

Weinert, F. E. (2001a). Concept of Competence: a Conceptual Clarification. In D. S. Rychen & L. H. Salganik, Defining and Selecting Key Competencies (pp. 45–65). Göttingen: Hogrefe & Huber.

Weinert, F. E. (2001b). Vergleichende Leistungsmessung in Schulen. In F. E. Weinert, Leistungsmessungen in Schulen - eine umstrittene Selbstverständlichkeit (pp. 17–31). Weinheim and Basel: Beltz.

Welzel, M., Haller, K., Bandiera, M., Koumaras, P., Nidderer, H., Paulsen, A., … von Aufschnaiter, S. (1998). Ziele, die Lehrende mit dem Experimentieren in der naturwissenschaftlichen Ausbildung verbinden - Ergebnisse einer europäischen Umfrage. Zeitschrift für Didaktik der Naturwissenschaften, 4(1), 29–44.

Weniger, E. (1952). Der Lehrplan. In E. Weniger, Didaktik als Bildungslehre. Teil I. Theorie der Bildungsinhalte und des Lehrplans (pp. 21–44). Weinheim: Beltz.

Westbury, I. (1995). Didaktik and Curriculum Theory: Are They the Two Sides of the Same Coin? In S. Hopmann & K. Riquarts (Eds.), Didaktik and/or Curriculum (pp. 233–264). Kiel: IPN.

Westbury, I. (2000). Teaching as a Reflective Practice: What Might Didaktik Teach Curriculum? In I. Westbury, S. Hopmann, & K. Riquarts (Eds.), Teaching as a Reflective Practice: The German Didaktik Tradition (pp. 15–54). Mahwah, NJ: Lawrence Erlbaum Associates.

Whitelegg, E., & Parry, M. (1999). Real-life contexts for learning physics: meanings, issues and practice. Physics Education, 34(2), 68–72.

Wicklein, R. C. (1993). Identifying Critical Issues and Problems in Technology Education Using a Modified-Delphi Technique. Journal of Technology Education, 5(1), 54–71.

Wimmer, M. (2003). Ruins of Bildung in a Knowledge Society: Commenting on the debate about the future of Bildung. Educational Philosophy and Theory, 35(2), 167–187.

Woudenberg, F. (1991). An evaluation of Delphi. Technological Forecasting and Social Change, 40(2), 131–150.

Yager, R. E. (1993). Science-Technology-Society As Reform. School Science and Mathematics, 93(3), 145–151.

Yang, Y. N. (2000). Convergence on the Guidelines for Designing a Web-Based Art-Teacher Education Curriculum: A Delphi Study. Paper Presented at the Annual Meeting of the American Educational Research Association April, 24-28, 2000, New Orleans, LA. Retrieved from http://eric.ed.gov/?id=ED446747 [15.06.2014]

Contents of Appendix

The contents of the appendix can be accessed via www.springer.com and "Theresa Schulte" within the OnlinePLUS programme.

Contents

Printed in the United States
By Bookmasters